THE TEN-THOUSAND YEAR FEVER

NEW FRONTIERS IN HISTORICAL ECOLOGY

Dynamic new research in the genuinely interdisciplinary field of historical ecology is flourishing in restoration and landscape ecology, geography, forestry and range management, park design, biology, cultural anthropology, and anthropological archaeology. Historical ecology corrects the flaws of previous ecosystems and disequilibrium paradigms by constructing transdisciplinary histories of landscapes and regions that recognize the significance of human activity and the power of all forms of knowledge. The preferred theoretical approach of younger scholars in many social and natural science disciplines, historical ecology is also being put into practice around the world by such organizations as the UNESCO. The series fosters the next generation of scholars offering a sophisticated grasp of human-environmental interrelationships. The series editors invite proposals for cutting edge books that break new ground in theory or in the practical application of the historical ecology paradigm to contemporary problems.

Series Titles

Vol. 1: *Social and Ecological History of the Pyrenees: State, Market, and Landscape,* Ismael Vaccaro and Oriol Beltran, eds.

Vol. 2: *The Ten-Thousand Year Fever: Rethinking Human and Wild-Primate Malarias,* Loretta A. Cormier

Vol. 3: *Sacred Geographies of Ancient Amazonia,* Denise P. Schaan

THE TEN-THOUSAND YEAR FEVER

Rethinking Human and Wild-Primate Malarias

Loretta A. Cormier

Left Coast Press Inc.

Walnut Creek, California

LEFT COAST PRESS, INC.
1630 North Main Street, #400
Walnut Creek, CA 94596
http://www.LCoastPress.com

ISBN 978-1-59874-482-8 hardcover
ISBN 978-1-59874-483-5 paperback
ISBN 978-1-61132-797-7 electronic

Library of Congress Cataloging-in-Publication Data:

Cormier, Loretta A.
The ten-thousand year fever : rethinking human and wild-primate malarias / Loretta A Cormier.
p. cm. – (New frontiers in historical ecology)
Summary: "Malaria is one of the oldest recorded diseases in human history, and its 10,000-year relationship to primates can teach us why it will be one of the most serious threats to humanity in the 21st century. In this pathbreaking book Loretta Cormier integrates a wide range of data from molecular biology, ethnoprimatology, epidemiology, ecology, anthropology, and other fields to reveal the intimate relationships between culture and environment that shape the trajectory of a parasite. She argues against the entrenched distinction between human and non-human malarias, using ethnoprimatology to develop a new understanding of cross-species exchange. She also shows how current human-environment interactions, including deforestation and development, create the potential for new forms of malaria to threaten human populations. This book is a model of interdisciplinary integration that will be essential reading in fields from anthropology and biology to public health"— Provided by publisher.

ISBN 978-1-59874-482-8 (hardback) — ISBN 978-1-59874-483-5 (paperback)

1. Malaria. 2. Medical parasitology. 3. Primates—Diseases. I. Title.
RA644.M2C587 2011
614.5'32–dc22

2011009013

Printed in the United States of America

⊗™ The paper used in this publication meets the minimum requirements of American National Standard for Information Sciences—Permanence of Paper for Printed Library Materials, ANSI/NISO Z39.48–1992.

Cover design by Piper Wallis
Cover illustration by James Cormier

Contents

Preface

My interest in studying human and nonhuman primate interactions (ethnoprimatology) began with my dissertation fieldwork with the Guajá, a group of Amazonian hunter-gatherers. The Guajá incorporate local primates into multiple aspects of their culture. Although monkeys are hunted and are a key food source, orphaned monkeys are nurtured as pets and considered to be quasi-human members of the kin group. Monkeys are also deeply embedded in Guajá cosmology, cosmogeny, and symbolic life. My book, *Kinship with Monkeys* (2003) was an ethnographic study that focused on how monkeys are intertwined in ecological, social, and religious dimensions of the Guajá culture.

Although the focus of my work with the Guajá did not specifically address disease ecology, I did observe several instances of human diseases being passed to pet monkeys. One case involved two young boys, who were cousins, who both developed what appeared to be *Herpes simplex*. Both had small cold sores that quickly resolved, but two of the pet monkeys in each of their households, a capuchin (*Cebus*) and howler (*Alouatta*) both became gravely ill, with lesions all over their faces. Both died within weeks. During my time with the Guajá, there were several instances when respiratory infections (likely due to a viral "cold" or "flu") spread through the group, and many of the pet monkeys experienced symptoms similar to those of the Guajá. It is also well known that multiple diseases can be passed from humans to wild primates and from wild primates to humans. Again, although it was not a focus of my research, I was aware that disease sharing was another important consequence of living in close quarters with primate cousins.

My specific interest in malaria developed from research in ethnobotany. However, for any ethnographer who works in malarious areas, malaria is an ever-present concern in the field. Many of the Guajá experienced serious malaria infections during my stay, and medical records from Brazilian health care workers have documented malaria cases since the time of first sustained contact with the Guajá people in the 1980s. The fear of malaria never leaves the mind, not only for the welfare of the people whom cultural anthropologists work among but also for one's personal health.

The primary purpose of my ethnobotanical research among the Guajá was to identify plants used for foods. One of the aims of my research was to quantify the relative importance of hunting primates in the overall diet. However, when I went out on plant-collecting treks, the Guajá repeatedly identified medicinals. I recorded that information but did not take a serious look at it until I left the field.

The Guajá have a class of medicinal plants that I have called "ghost repellents." The Guajá believe that lonely ghosts of the dead (the *aiyã*) cannibalize the souls of the living and create illness. One of the key symptoms of such "ghost infection" is fever. The Guajá bathe in infusions of pungent-smelling plants that are thought to ward off the spirits of the dead. They relate that the ghosts of the dead are foul-smelling, like corpses, and that the odor of ghosts is one mechanism through which they can permeate the body and invade the soul. Bathing in strong-smelling plants can provide a protective barrier against the *aiyã*.

I initially believed that this practice could be relegated to placebo effect, being primarily part of their religious belief system. However, in researching the ethnomedicine of other Amazonian groups, I found a similar practice among a number of peoples who were linguistically and geographically distant. This situation suggested to me that the bathing in aromatic plants might be a very ancient practice in Amazonia and might be an attempt to combat a specific disease. Since the key symptom of ghost infection is fever, malaria seemed a likely culprit. Although most researchers believed malaria to have been introduced to the New World at the time of colonization, I was aware that a few had hypothesized that human malaria was present in the New World before contact. Thus the initial hypothesis that I began investigating was that Amazonian antifever/antighost plants were an ancient attempt to ward off malaria.

To make a long story short for the purposes of this preface, the bulk of the evidence currently available seems to lean far more heavily on the side that malaria was not present in either human or wild-primate populations in the New World before colonization. But as is often the case in research, we often learn far more from being wrong about our initial assumptions than from being right. What was initially intended as a short and straightforward article on the relationship between aromatic plants and malaria became more and more complicated the more I tried to answer the question. I wound up enrolling in several public health courses in tropical infectious disease, epidemiology, and global health and quickly realized that the "Amazonian" question could not be answered by reference to Amazonia alone. I found multiple puzzles in the relationship between human and wild-primate malarias, which this work only partially resolves.

This book is more timely than I originally intended or imagined. I started this research before *P. knowlesi* was deemed "fifth human malaria." *P. knowlesi* is a wild-primate malaria that was recently identified to be prevalent in human populations in Southeast Asia. Its discovery in human populations engendered a spate of new research into human and wild-primate malaria transfers. Historically, it is reminiscent of a similar interest in human-wild-primate connections in 1960 when its researchers found, to their amazement at the time, that humans could contract wild-primate *P. cynomolgi*.

Although this work is timely, the downside of writing a book that attempts to synthesize the literature to date in an area where intensive research is focused is that it can provide up-to-date findings only as of the date of publication. Perhaps most important is the relationship of human falciparum malaria to related wild-primate malarias. When I first began writing this book, the only two major competing hypotheses regarding the origins of falciparum malaria in humans were an older hypothesis that it was a zoonotic transfer from birds at the time of the origin of agriculture (10,000 years ago) and the more widely accepted newer hypothesis that falciparum-type malaria existed in the African great apes and diverged into a human *P. falciparum* line and a great ape *P. reichenowi* line 5 million years ago.

I have rewritten the falciparum chapter several times since the first draft of this book—first incorporating new studies in molecular genetics indicating that falciparum may have been a recent zoonotic transfer from chimpanzees, and then incorporating findings that Western gorillas and bonobos may be either reservoirs or the source of human falciparum malaria. Important new findings have also been uncovered recently in anthropology, including the new Flores hominin, Denisovan hominin, and many new *Homo erectus* findings in Asia. I have done my best to make the final version of this book as up-to-date as possible. But given that the ethnoprimatology of malaria is such an active field of research, my latest synthesis may prove to be outdated by new findings. So, this work should be read as a best effort to incorporate relevant findings across disciplines.

MALARIA

The 10,000 Year Fever approaches malaria from a historical ecological perspective. Malaria is a manifestation of complex co-evolutionary relationships among hosts, vectors, and parasites that have been in play for at least 30 million years. However, the status of malaria as an *epidemic* human disease likely can be traced to the origin of agriculture, approximately 10,000 years ago. As such, it should be considered an inherently biocultural phenomenon, because human modification of the landscape has amplified malaria and resulted in multiple, dynamic, and ongoing biological and cultural adaptations. One aspect of this complex that is often neglected is the role of nonhuman primates. From prehistory to present, cross-species infections of malarias have interdigitated and affected the disease ecologies of humans, apes, and monkeys. A key aim of this book is to integrate ethnoprimatology into the understanding of the historical ecology of malaria.

Malaria has the dubious distinction of being one of the oldest recorded diseases in human history and also of remaining one of most serious health threats to humanity in the 21st century. It is described in the earliest known historical records of the ancient physicians of Greece, Rome, India, and China. Today malaria rivals AIDS in terms of mortality and exceeds AIDS in terms of morbidity. In 2007 the WHO estimated that 2.1 million people died from AIDS, whereas 1–3 million died from malaria (estimates for deaths from malaria are less precisely known). Although approximately 33 million people were living with HIV/AIDS in 2007, 300–500 million cases of malaria were recorded. Thus, as far as prevalence is concerned, malaria outpaces HIV/AIDS at least 10-fold.

Malaria has a long evolutionary history in the order Primates. However, the title of the book, *The 10,000 Year Fever*, points to a change in the historical trajectory of malaria whereby culture, ecology, and biology intersect. Human modification of the landscape through swidden agriculture created new mosquito habitats that fueled malaria. Selective pressures resulted in human biological adaptations providing some malarial resistance but also gave rise to genetic disorders in human populations such as sickle-cell anemia and thalassemia.

The intensification of malaria has led to multiple cultural adaptations and cultural consequences. It has affected human settlement patterns and seasonal migrations and has shaped human social organization both in terms of cross-cultural relationships among peoples of a region and intracultural dynamics, including class relations, gender roles, and racially based hierarchies. Antimalarials appear in the ethnobotanical inventories of many indigenous peoples in tropical and temperate zones. Malaria has also influenced symbolic systems and has a role in the ritual practice and supernatural beliefs of many cultures. On a larger scale, European colonization and the African slave trade introduced malaria into new populations, with devastating effects for New World indigenous peoples. In North America, colonial agricultural practices simulated malaria intensification in prehistoric Africa. Old World populations both introduced plasmodia and recreated ecological conditions on a new continent that fostered mosquito proliferation. In South America, malarial epidemics have appeared in the wake of deforestation and development from colonial times to today.

The genetic relationships among the plasmodia affecting human and nonhuman primates suggest that malaria is best understood as four separate disorders affecting anthropoid primates as a whole. It is understandable, from a public health perspective, to make a distinction between the plasmodia identified as causative for "human" versus "nonhuman" malarias. However, in terms of genetic phylogeny, such a dichotomy is false. Four plasmodia are responsible for human malaria: *P. falciparum*, *P. vivax*, *P. malariae*, and *P. ovale*. These four "human" malarias are only distantly related to one another. With the exception of *P. ovale*, each type of human malaria has been demonstrated to have far closer genetic relatives in wild-primate forms. For example, the human *P. falciparum* is related to African ape *P. reichenowi*. Human *P. vivax* malaria is closely related to approximately one dozen plasmodia-causing malarias in Southeast Asian primates. *P. malariae* is considered to be genetically indistinguishable from New World monkey *P. brasilianum*. Thus, there are four separate lines of evolutionary relationships among the anthropoid primate malarias rather than there being a genetically valid basis for distinguishing between human and nonhuman primate malaria forms.

The genetic relationships of the plasmodia reveal a complicated picture of cross-infections of malarias among human and nonhuman primates over time. These malarias cannot be considered to be either zoonoses or anthroponoses in the conventional sense, because the plasmodia are neither host-specific nor specific even at the level of primate infraorders, given that the relatively distant platyrrhine monkeys and humans are infected with the same parasite. The malarias cannot be considered to be geographically distinct, because they have crossed the continental divides of Africa, Eurasia, and the Americas. What is apparently more relevant for a given plasmodium to take hold is consideration of the local ecological conditions that provide efficient vectors able to make contact with susceptible primate hosts.

Humans, nonhuman primates, and mosquitoes within a given region adapt to differing niches, which may or may not overlap. A number of factors can bring human and nonhuman primates into contact with the same vectors. Indigenous peoples and wild primates often exploit the same forest space, and both share water sources with mosquitoes in the tropical forests. Many Amazonian peoples hunt monkeys and keep them as pets, bringing them into direct contact with one another. Globally, deforestation and development have led to human encroachment into nonhuman primate habitats.

Deforestation and development have the potential to foster new cross-species exchanges of malaria. Such a scenario apparently occurred recently in Borneo with a large-scale outbreak of macaque *P. knowlesi* among humans in several areas. Several malariologists have dubbed *P. knowlesi* as the "fifth human malaria" and consider it a serious threat as a new, emerging human disease. However, the extent to which nonhuman primate malarias affect humans (and vice versa) may not be known, because basic laboratory tests do not distinguish well among closely related malarias. In Borneo, the *P. knowlesi* outbreak was initially diagnosed by laboratory smear as human *P. malariae*. More generally, cross-species infections have important implications for malaria eradication programs, because it is insufficient to combat malaria as a human disease if it is being maintained in nonhuman primate reservoirs. On a very broad scale, the absence of nonhuman primate reservoirs may very well have been one factor in the ability to eliminate malaria in North America and Europe while it has remained entrenched in Africa and South America.

Malaria research is perhaps one of the most interdisciplinary fields of inquiry, involving complex interactions in the ecologies, biologies, and behaviors of parasites, vectors, and hosts. Contemporary research has included fields as diverse as geneticists mapping plasmodia genomes, entomologists studying mosquito biology and behavior, anthropologists studying human cultural and biological adaptations, and health professionals focusing on prevention and treatment. Thus malaria is particularly amenable to a historical ecological research program, not only because of its multidisciplinary nature but also because of its emphasis on landscape as a human-environmental interactive sphere, transforming over time. In Balée's (2006:77) recent review article of historical ecology, "landscape" can be understood as "a place of interaction with a temporal dimension that is as historical and cultural as it is evolutionary per se, if not more so, upon which past events have been inscribed, sometimes subtly on the land." Although landscape can also be conceptualized as situating human behavior and agency in the environment, several social theorists (for example, Kohn 2007, Latour 2004, Rabinow 1996) have called for social and ecological frameworks that involve transpecies actors. A key aim of this work is that the landscape of malaria is best approached as a cross-primate/environmental interactive sphere.

Neither the history of the human malarias nor the wild-primate malarias can be fully understood without considering that plasmodia opportunistically host-switch among the anthropoid primate species. Although this work demonstrates that cross-infection has been an ongoing process throughout anthropoid primate evolution, it is particularly relevant today. Deforestation and development are creating new zones of primate interaction, with both human and wild primates encroaching into each other's habitats. Numerous recent studies portend increased lateral transfer of plasmodia among the primates. Given that trend, the potential exists for novel forms of malaria to become established in primate hosts, emerging as new health threats to both human and wild-primate populations.

CHAPTER OVERVIEW

Chapter 1, *Malaria as a Primate Disorder*, makes the case for the relevance of approaching malaria as a generalized primate disorder. The evolutionary history of the plasmodia that cause malaria is older than the emergence of the primates, and even older than the emergence of mammals. The plasmodia responsible for malaria today are descendents of ancient protozoa that parasitized the intestinal tract of the common ancestor of birds, reptiles, and amphibians. Malaria has been identified as occurring in over 170 contemporary species. Of these, more than half afflict primate species. In terms of the distribution of malaria, primate species stand out as being particularly vulnerable.

The phylogenetic relationships of the plasmodia that infect primates are described, including recent advances in molecular genetics that have shed light on the close relationship among the primate malarias. Although four species of plasmodia cause malaria in humans, the relationships of the human species are far more closely related to plasmodia that affect nonhuman primates than they are to one another, with some human and nonhuman primate plasmodia being genetically indistinguishable.

A broad classification is proposed for the primate malarias, (1) falciparum-type malaria, (2) vivax-type malaria, (3) malariae-type malaria, and (4) ovale-type malarias, which are used throughout the book. Analysis of the genetic relationships of the primate malarias reveals a complex history of human and wild-primate cross-infections not only in the distant past but also in cases of recent plasmodial host-switching. The argument is made that these cross-infections cannot be considered to be either zoonoses or anthroponoses in the conventional sense, but rather that the appropriate level of analysis is that of the anthropoid primates.

Chapter 2, *Co-Evolution: Parasites, Vectors, and Hosts*, involves a discussion of the biology and adaptation to malaria, co-evolutionary and historical ecological and co-evolutionary processes, and problems with assumptions that levels of virulence provide a valid means of evaluating the time-depth of the length of association between primates and the four plasmodia types. First, an overview is provided of the life cycles of the plasmodial parasites, mosquito vectors, and primate hosts. Evidence exists for biological adaptation to malaria in both human and nonhuman primates. The best known adaptations are human polymorphisms that violate Hardy-Weinberg equilibrium such as sickle-cell trait, RBC Duffy negativity, thalassemia, and G6PD. Although far less is known about potential nonhuman primate polymorphisms, the evidence is reviewed for blood phenotypes suspected to provide resistance in several great ape species.

Considerable debate exists in the literature regarding whether or not vivax-type malarias should be considered to have had a longer evolutionary history in primates than falciparum-type malarias, because (1) vivax-type malarias are relatively benign in comparison to falciparum-type malarias in humans, and (2) in West and Central Africa, RBC Duffy negativity confers complete immunity to vivax-type malaria in humans, whereas adaptations such as sickle-cell trait confer only partial resistance to falciparum malaria. The implication is that vivax is older, because more efficient biological adaptations have evolved and co-evolutionary processes may have led to decreased virulence for vivax-type malarias.

While numerous examples exist of host-pathogen co-evolution that results in increased resistance in hosts and decreased virulence in pathogens, this is only one co-evolutionary process. Increased virulence is selected for in some host-pathogen relationships. The limitations of the virulence model are discussed, and attention is given to how co-evolutionary processes are more complicated in malaria and other vector-borne diseases, because three organisms are involved. Attention is given to the possibility that the mosquito-pathogen relationship may be more important to understand co-evolutionary processes in malaria than the human-pathogen relationship is. In the case of human falciparum malaria, increased host virulence enhances the survival of the parasite. The survival rates of mosquito vectors are also reduced, since carriers are sluggish, bite more often, and engage in riskier biting behavior. For vivax malaria, the sporozoites can enter a dormant phase, which normally lasts several years in tropical and temperate climates, where malaria is prevalent today. But in Scandinavia in the 18th century, the hypnozoite stage extended nearly a decade, and malaria epidemics have reached as far as Siberia.

Chapter 3, *Falciparum Type: The Great Ape Malaria*, provides an overview of the revolution in understanding of the historical phylogeny of malaria owing to recent advances in molecular genetics. A description is provided of the advances and limitations of methods used in phylogenetic reconstruction, including the current state of mapping of the *P. falciparum* genome. Given that *P. falciparum* is the deadliest of the human malarias, as discussed in Chapter 2, conventional wisdom had long held that it was of more recent origin than other more benign forms of malaria. Recent work in molecular genetics have upended the belief that only two forms of falciparum-type malaria existed in wild primates: human *P. falciparum* and African ape *P. reichenowi*. Several new forms of falciparum-type malaria have been identified in African apes, and it has also been discovered that they harbor *P. falciparum*.

A "Malaria's Eve" hypothesis has been promulgated that suggests that human *P. falciparum* s.l. underwent a bottleneck effect giving rise to a *P. falciparum* s.s. that represents all extant forms of malaria in human populations. Malaria's Eve is consistent with the agricultural hypothesis first proposed by Frank Livingstone in 1958. He argued that iron tools and swidden agriculture led to anopheline vector proliferation and the development of sickle-cell trait as an adaptation to falciparum malaria in tropical Africa. Although some advocates of the agricultural hypothesis erroneously supposed that falciparum malaria was acquired zoonotically from birds, broadly, the genetic evidence of Malaria's Eve is consistent with the cultural evidence, which together suggest a big bang of falciparum malaria in human populations coincident with the advent of agriculture in Africa. Moreover, although a "big bang" of agriculture in Africa is a convenient heuristic device, it is important to recognize that agriculture did not emerge *sui generis* but gradually developed from paracultivation to intensification with iron age technology. Landscapes favorable to malarial vectors emerged gradually as did settlement patterns and population densities that fostered proliferation.

Chapter 4, *Vivax Type: The Macaque Malaria*, discusses the debates surrounding the origin and spread of vivax-type malarias in both the Old World. Two competing hypotheses have been proposed: Out of Africa and Out of Asia. The Out of Africa hypothesis argues for the great antiquity of vivax malaria, perhaps older than falciparum malaria. Two arguments bolstering this view are (1) the high prevalence (~95%) of RBC Duffy negativity in tropical Africa, which confers complete immunity to vivax malaria, suggesting a long period of co-evolution. Similarly, (2) vivax is suggested to be older than falciparum because it manifests itself as a less virulent form of malaria. In this view, modern human populations brought vivax malaria into Southeast Asia, where it was introduced to nonhuman primates by way of lateral transfer. The Out of Asia hypothesis argues that vivax malaria originated in situ among Southeast Asian wild primates and was laterally transferred to humans. The key argument is that the great diversity of vivax-type malarias among Southeastern Asian wild primates suggests a center of endemicity.

Some researchers believe that Pleistocene refugia isolated vivax malaria (in either human populations or wild-primate populations), requiring a recent lateral transfer. In this chapter, a third potential scenario is proposed based on recent revisions to our understanding of hominoid evolution in Asia. Namely, *P. vivax* may have had a continuous presence in hominoids in Asia. It could have entered at any point from the Miocene apes to *Homo erectus* or to modern *Homo sapiens*. Recent fossil discoveries of *Homo erectus* have made a scenario of a continuous hominoid presence feasible since Miocene times. Names, the evidence suggests the presence of *Homo erectus* in Asia both earlier and later than previously known. In Southeast Asia, *Homo erectus* was present as early as 2 million years ago and as late as 50,000 years ago. Although the dates remain controversial, some argue that *Homo erectus* was present in Indonesia as late as 27,000 years ago, and some consider the newly discovered Flores hominin (12,000 B.P.) to be a relict form of *Homo erectus*. Clear evidence exists for the presence of *Homo sapiens* in Malaysia by 40,000 B.P. Thus, even by conservative estimates, island Southeast Asia can be characterized as having a continuous hominin occupation from approximately 2 million years ago to the present. From the Miocene onward, a continuous fossil record exists of hominoid primate presence in Asia. This suggests that isolating Pleistocene refuge areas may not be essential to the debate. Although it is possible that a recent lateral transfer of vivax malaria occurred between modern humans and wild Southeast Asian primates, it is also possible that vivax malaria has long existed as a generalized disease of anthropoid primates, or, that a lateral transfer could have occurred at any point from the Miocene to recent times. Contemporary vivax malaria in primates may also represent a recent bottlenecking, similar to Malaria's Eve hypothesis for falciparum malaria.

Chapter 5, *Migration: Malaria in the New World*, discusses a recent puzzling finding from molecular genetics concerning the relationship among human and wild-primate vivax-type malarias that has led some researchers to suggest that vivax malaria originated in South America. The subspecies of vivax-type malaria present in New World monkeys is more closely related to vivax-type malarias in Southeast Asian wild primates than it is to human vivax malaria in South America. It has long been assumed that all forms of New World malaria are virgin soil diseases, introduced with colonization. If the closest relative of New World monkey vivax malaria is found to be Southeast Asian wild-primate vivax malaria, that indicates that human and nonhuman primate vivax malarias in South America derive from two separate origins.

Malarial geneticists have proposed two hypotheses to account for New World monkey vivax malaria. One hypothesis suggests that vivax malaria originated in situ in South American primates and spread from South America to the Old World as recently as 500 years ago, following colonization. More weight has been given to the alternative suggestion of a Pre-Columbian introduction of vivax-type malaria. Here, geneticists have discounted the possibility of an Amerindian lateral transfer to South American primates due to both the improbability of plasmodia surviving the cold barrier of the Arctic and the logistics of essentially needing to migrate across the expanse of three continents in a relatively short period of time for humans to transfer malaria to South American monkeys. What has been suggested is that Pre-Columbian peoples in Southeast Asia traveled by boat to South America and laterally transferred vivax-type malaria to New World monkeys. Later, with colonization, another form of vivax-type malaria was introduced to South American Indians from European peoples. I argue that although recent archaeological evidence suggests Pre-Columbian contact between Remote Oceanic populations and South American populations, doubling the distance of boat travel to Southeast Asian peoples is most unlikely. The hypothesis is proposed that Amerindian vivax malaria derived from a lateral transfer from European vivax malaria and that simian vivax malaria derived from a much later lateral transfer from Asian migrant workers in Brazil in the 19th century.

Chapter 6, *Rhesus Factor: Experimental Studies in Wild Primates,* and Chapter 7, *Ethics: Human Experimentation,* describes the history of experimental studies involving attempts to cross-infect human and wild-primate malarias under laboratory conditions. The use of nonhuman primates in biomedical research for malaria did not originally derive from an understanding of the close relationship of the primate malarias but rather from historical factors that led to increased reliance on rhesus monkeys as the gold standard in biomedical research beginning with polio studies in the 1940s and solidifying with the thalidomide studies and Harry Harlow's psychological experiments in the late 1950s and early 1960s. Chapter 6 describes numerous studies that have involved experimental infection of monkeys with various human and nonhuman forms of malaria. The results reveal varied responses to exposure to malaria, including natural resistance in species, virulent reactions, and within-species individual differences. These findings are evaluated in light of the hypothesis of differential susceptibility to vivax-type malaria driving speciation and geographic distribution of Old World primates in Southeast Asia.

Although endemic malaria had been effectively eradicated in the United States by mid-century, cases of malaria in U.S. citizens returning from travel abroad raised the specter of the possibility of malaria becoming reestablished on U.S. soil. The World War II Stateville prison malaria studies are well documented, but less attention has been given to the Atlanta Penitentiary studies of the 1960s. The Stateville malaria project involved infecting prisoners with human malaria and testing antimalarial drugs. The Atlanta Penitentiary studies involved intentional exposure of inmates with wild-primate malarias to determine whether infections could be established. Before 1960, it was not believed possible for humans to acquire a nonhuman primate malaria until a researcher was accidentally infected with *Plasmodium cynomolgi* (common to macaques). Articles in the *American Journal of Tropical Medicine and Hygiene* reveal that over a 10-year period, several hundred prisoners were inoculated with various strains of wild-primate malaria, with success in transmitting the disease for most strains. Wild-primate malaria was also used for a short time as a treatment for neurosyphilis. Also, beginning in the 1960s, experimental studies were initiated that attempted to induce various types of nonhuman primate malaria in various novel nonhuman primate hosts. The recent outbreaks of macaque *P. knowlesi* in Indonesia are addressed in the final chapter.

Chapter 8, *Future: The Primate Malaria Landscape*, discusses the ways in which human modification of the landscape has affected the proliferation of malaria. Although malaria is a disease of great antiquity in primates, the genesis of malaria as an epidemic disease in humans likely began with the advent of agriculture. In Africa, swidden agriculture created new habitats for mosquitoes that amplified malaria and resulted in multiple, dynamic, and ongoing biological and cultural adaptations. European colonization and the slave trade in North and South America introduced malarial parasites and agricultural practices, which encouraged the spread of malaria into new ecological areas. Not only do humans and nonhuman primates share malaria vectors, but many nonhuman primates opportunistically exploit agricultural fields as crop raiders in both the Old and New World. Fallow fields provide an overlapping environment for host succession from human to wild primates, being ideal habitats for mosquito breeding and also attracting many primate species to plants thriving in secondary forest habitats. Deforestation and development have led to human encroachment into wild-primate habitats and wild-primate encroachment into human habitats, also increasing the likelihood of disease transfer.

This chapter also discusses the probable underestimation of the extent of human-nonhuman primate cross-transfer of *P. malariae* in human populations. WHO statistics estimate that *P. malariae* occurs at a rate of <1% in the countries included in Amazonia. However, numerous studies with Amerindian groups that hunt monkeys and keep monkeys as pets have described *P. malariae* positive smears at rates as high as 100% among adults. These findings disappear statistically for numerous reasons. One is that Amerindians represent a minority population in Amazonia, so their unique conditions are not well represented in aggregate WHO statistics. Another is that in Brazil and in other nations included in Amazonia by WHO, smears for malaria are initiated only if individuals seek health care for malaria symptoms. It is necessary for Amerindians both to have regular access to health care and for symptoms to be severe to seek out Western medical intervention. Since *P. malariae* is a relatively less virulent form of malaria and the vast majority of Amerindian peoples have limited access to Western health care, it is likely that the prevalence of *P. malariae* has been grossly underestimated among indigenous peoples.

The WHO and other allied organizations have recently advocated the reinstitution of DDT to control malaria. Although DDT is indeed an effective measure to exterminate mosquito vectors locally, it fails to consider the potential problem of wild-primate reservoirs of malaria in the tropics. DDT may have played an important role in eradicating malaria in the United States and Europe, but these areas did not harbor wild primates (except for the small Barbary macaque population in Gibraltar). Ongoing development and deforestation are fostering increased contact between human and wild-primate populations, facilitating lateral transfers of plasmodia. Any program for malaria eradication must address cross-infection between humans and wild primates, because the evidence suggests that prehistorically, historically, and in contemporary times, host-switching is characteristic of the plasmodia responsible for the primate malarias.

Over the last decade, cross-species malaria infections have been increasingly documented under natural conditions, not only between humans and wild primates but also among wild-primate species. In 2004 a large-scale outbreak of wild primate *P. knowlesi* occurred in humans in Borneo, with several similar cases appearing in the Philippines in early 2008. What is unknown is if *P. knowlesi* represents a newly emergent form of malaria in human populations or if it is already well established and only recently discovered. In South America, evidence exists for increasing rates of human/monkey *P. malariae/P. brasilianum* cross-infections in French Guiana. One recent primate survey in Brazil found several cases of *P. falciparum* antibodies in wild primates, indicating anthroponotic infection. Given that many New World monkeys are very susceptible to falciparum malaria under experimental conditions, it raises the possibility of a frightening scenario if falciparum malaria were to become established in wild primates in Amazonia. The recent discovery of *P. falciparum* in bonobos, chimpanzees, and gorillas may indicate that African apes have long served as a reservoir for *P. falciparum*, complicating the ability to control that particularly virulent form.

Over the last 20 years, several new emerging diseases have been either determined to have a wild-primate link or have been suspected to have one. Perhaps the best known has been the demonstration 10 years ago that human HIV is a zoonotic infection derived from the common chimpanzee (*Pan troglodytes troglodytes*). Members of the ebola filovirus group can infect both human beings and wild primates. The natural hosts of the ebola viruses remain unclear; several are known to be capable of being spread among primate species. Other viral hemorrhagic fevers, such as the Mapucho and Lassa viruses, infect both human and wild primates. One cautionary tale is yellow fever. An extremely effective vaccine exists for yellow fever, yet the fever has not been eradicated in South America and Africa, in part, because it is entrenched in monkey populations, and an enzootic cycle persists with continuous cross-infections between monkeys and unvaccinated humans. The focus of this work, the landscape of malaria, includes historical changes in habitats, human and wild-primate interactions, and increasing cross-infections. Its relevance extends beyond malaria itself in providing a potential framework for approaching both past and future health threats to human and wild-primate populations.

Acknowledgments

I would like to acknowledge the advice and support of many people who have contributed to the successful completion of this book. Although I would like to recognize those who have been helpful to me, this work is my own. Whatever its strengths and weaknesses, they are all my responsibility alone. First I would like to thank the Guajá people, who graciously accepted an outsider into their midst, shared their lives with me, and taught me how to be an anthropologist. I would also like to thank the following people who provided advice and comments in an early article draft on Neotropical primate malaria, which was the starting point of this book: Pauline Jolly (Department of Epidemiology, UAB), Julian Rayner (Division of Geographic Medicine, UAB), and Oralee Branch (Division of Geographic Medicine, UAB). That portion of this research was also supported by a University of Alabama at Birmingham Faculty Development Grant. I would also like to thank Left Coast Press, including Jennifer Collier, the Historical Ecology Series editors, Bill Balée and Carole Crumley, and the anonymous reviewers. I would also like to acknowledge a number of students at UAB who assisted in retrieving articles, providing editorial comments, indexing, and so on: Caitlin Aamodt, Susie Caflin, Jade Delisle, Anna McCown, Mallory Messersmith, and Ashley Wilson. I would like to say a special thanks to three friends and colleagues at UAB who provided support and encouragement, Lisa Baker, Sharyn Jones, and Bruce Wheatley (who was also my first primatology professor). Finally, I would like to thank my dear husband James for designing the cover of this book and being my best friend and greatest supporter for the last 25 years.

Loretta A. Cormier
May 2011

Chapter 1

Introduction: Malaria as a Primate Disorder

In this opening chapter, a case is made for the relevance of approaching malaria at the level of the primates. A broad overview of some of the key issues is provided here and is elaborated on in subsequent chapters. The phylogenetic relationships of the plasmodia infecting primates are described, including how recent advances in molecular genetics are changing our understanding of the relationship among the primate malarias. Humans are natural hosts to four species of malaria-causing plasmodia: *Plasmodium falciparum, P. vivax, P. ovale,* and *P. malariae.*[1] These four parasites are not closely related to one another and separately colonized hominins[2] (humans and their ancestors). However, each of the four species is closely related to other plasmodia that infect nonhuman primates. For example, human *P. falciparum* and *P. vivax* are not closely related, but human *P. falciparum* is closely related to several great ape parasites, and *P. vivax* is closely related to several Old World monkey parasites.

Approximately 40 different species of malaria parasites have been identified in primates (see Appendix 1 for references). To date, genetic relationships have been established among about half of them, which can be divided into four main groups among the anthropoid[3] primates. However, within these types, many of the plasmodia species are non-specific according to particular host or have a history of host-switching across species. The malaria parasites as a whole are older than the emergence of the primates and even older than the emergence of mammals. The plasmodia responsible for malaria today are believed to be descendents of ancient protozoa (endocellular coccidian parasites) of the intestinal tract of the common ancestor of birds, reptiles, and amphibians (Cogswell 2000). The plasmodia affecting primates are differentiated here into four main groups: falciparum type, vivax type, malariae type, and ovale type. Analyses of the genetic relationships among the primate malaria parasites have revealed a complex history of human and wild-primate cross-infections, not only in the distant past but also in cases of relatively recent plasmodial host-switching. The terms "zoonosis" (transmission of disease from a nonhuman animal to a human) and "anthroponosis" (transmission of disease from a human to a nonhuman animal) are used in this work. However, arguably, cross-infections cannot be considered to be either zoonoses or anthroponoses in the conventional sense; rather anthropoid primates may be the most appropriate level of analysis for understanding the primate malarias.

Overview of Primate Malaria

A disease consistent with the symptoms and ecology of malaria is recorded in numerous medical texts of the ancient world, including the writings of Greek, Roman, Ayurvedic, and Arabic physicians. In the 4th century B.C.E., frequent references are found in the works of the Greek physician Hippocrates describing an illness with symptoms that seem to suggest malaria, including cycling fevers, splenomegaly, and a relationship of the illness to habitats of low, marshy areas (Bruce-Chwatt and de Zulueta 1980). Similar descriptions of an illness characterized by periodic fevers can also be found in the works of Celsus, Pliny the Elder, and Galen (Dunn 1993) as well as in the works of Classical Roman physicians such as Martialis, Juvenal, Horace, and Livy (Stannard 1993). In South Asia, the Ayurvedic Hindu medical text Susruta (by 6–7th century C.E.) also describes a febrile disease with a relationship to mosquitoes that likely refers to a form of malaria (Chakravorty 1993). Arab physicians in the 11th and 12th centuries in Islamic Spain describe cyclical fevers linked to stagnant water sources (Bruce-Chwatt and de Zulueta 1980).

The earliest typology of malaria was based on the periodicity of the fever cycle. Beginning with Hippocrates, a distinction was made between "tertian" and "quartan." Tertian malarias are characterized by occurring every third day, and quartan malarias occur every fourth day. In the ancient Greek description, both the day of onset and the day of crisis were included, so tertian cycles are actually 48 hours and quartan are 72 hours (Stannard 1993). These terms are still in use today with *Plasmodium vivax, P. ovale,* and *P. malariae* described as tertian malaria and *P. falciparum* described as quartan malaria. When malaria began to be documented in wild primates, early typologies also relied on periodicity for classification (Coatney et al. 1971). However, it is now known that such life history features can be plastic and are not a reliable way to determine the relationships among plasmodial forms. Human malarias are also classified frequently in contemporary research according to the severity of symptoms, contrasting "benign" malaria (*P. vivax, P. ovale,* and *P. malariae*) with "malignant" or "severe" malaria (*P. falciparum*).

Although we may very well have a historical record of malaria going back to the time of Hippocrates in the 4th century B.C.E., the causative agent for malaria was discovered less than 150 years ago in 1880 by the military physician Alphonse Laveran, which he named *Oscillaria malariae* (Sherman 2005). The transmission of malaria by mosquitoes was discovered by Ronald Ross, a surgeon-major in the Indian medical service, who published his finding in 1897 (Sherman 2005). The genetic relationships among the primate malarias have begun to be elucidated only within the last 25 years.

Historically, it has been common in the literature to make a distinction between the human and the nonhuman primate (or simian) malarias. From a public health perspective, the separation is logical, because the primary focus is in understanding the malaria parasites that have been identified as significant agents of human disease. Such a division can be problematic, because it prioritizes the host over the parasite. Some species of plasmodia infect both human and nonhuman primates. Also, the division of primates into the categories of human and nonhuman is not taxonomically valid for the order Primates. For example, humans, bonobos, chimpanzees, gorillas, and orangutans (hominids) are far more closely related to one another than any are to New World monkeys. However, in this text, I do frequently make the distinction between human and nonhuman (or wild) primates. The rationale is more anthropological than public health oriented, in making our traditional distinction in gradistic differences among the primates, particularly regarding relevance of factors related to the elaboration of human culture. Nonetheless, such a distinction is expository rather than taxonomic.

As previously stated, four species of plasmodia are responsible for human malaria infections: *P. falciparum, P. vivax, P. malariae,* and *P. ovale*. Genetically, these four plasmodia are not closely related, with *P. falciparum* being the most distant. Each of the four has undergone separate evolutionary histories and has closer genetic relatives among the wild-primate parasites than any have to one another. For that reason, the primate malarias will be distinguished as four broad types that manifest across human and nonhuman primate species: falciparum type, vivax type, malariae type, and ovale type. This typology is consistent with the basic classification scheme developed by the first synthetic overview of the primate malarias (Coatney et al. 1971). However, in the original work of Coatney and associates (1971), classification was based on gross morphology and periodicity, which has proven to have been misleading in light of recent advances in the molecular phylogeny of plasmodia. For example, *P. fragile* was judged to be closely related to *P. falciparum,* and *P. simiovale* was thought to be closely related to *P. ovale* (hence the now erroneous name "simiovale"), but both *P. fragile* and *P. simiovale* are now known to be more closely related to human *P. vivax* than to *P. falciparum*.

Less than two years ago only two falciparum-type primate malarias had been identified: human *P. falciparum* and African ape *P. reichenowi*. The picture has become much more complicated now in that at least three new falciparum-type lines have been identified in chimpanzees (*P. billcollensi, P. billbrayi,* and *P. gaboni*) and two in gorillas (*P. GorA* and *P. GorB*). In addition, *P. falciparum* has been identified in bonobos, chimpanzees, and gorillas. The vivax type includes human *P. vivax* and at least 10 related forms that infect New World monkeys, Old World monkeys, and Asian apes. The malariae type includes human *P. malariae,* New World monkey *P. brasilianum* (affecting at least 35 species), and chimpanzee *P. rodhaini*, which may be the same parasite as *P. malariae*. The ovale type includes human *P. ovale* and possibly *P. schwetzi*.[4] Very recently, *P. ovale* and has been found in chimpanzees in Cameroon (Duval et al. 2009, 2010) which the researchers believe to represent natural cross-infections.

Primates appear to be particularly vulnerable to the malaria parasites. Sinden and Gilles (2002) estimate that malaria parasites infect approximately 120 living species of mammals, reptiles, and birds. A review of the literature (see Appendix 1) found malaria parasites documented in 65 primate species; thus species infecting primates represent half of all identified malaria parasites. As is described in depth in this work, within the four major types of plasmodia specific to primates, some have demonstrated an ability to laterally transfer, or "host-switch," across primate species, genera, and even families. Others appear to be more host-specific.

I propose that the most appropriate grouping for the four types of parasites (falciparum, vivax, ovale, and malariae) is as *primate malaria* and that the level of anthropoid primate is the most appropriate level of analysis for understanding these four types of plasmodia. At this time, the grouping can be applied only to the anthropoid primates (monkeys, apes, and humans). The prosimians remain an open question. These so-called "lower primates" are small-bodied, primarily nocturnal species that include tarsiers, lemurs, lorises, galagos, and bushbabies.[5] Although eight species of malaria parasites have been identified in prosmians (see references in Appendix 1), insufficient research has been done at this time to be able to determine their genetic relationships to the anthropoid primate plasmodia.

The four types of parasites appear to be specific to anthropoid primates. Thus far, we do not have evidence that any of these plasmodia naturally parasitize taxa outside the primates. However, within the primates, species within each of the four types infect a variety of primate species. Many of the specific plasmodia species infect more than one species of primate, some of which are as distantly related as humans and platyrrhine (New World) monkeys. In addition, many primate species are infected by more than one species of plasmodia. Most of these multiple infections occur within the vivax group in Asian macaques. However, some New World monkeys are infected with both malariae and vivax type, chimpanzees have demonstrated mixed infections of falciparum and malariae type (and also the unclassified *P. schwetzi*), and humans can also have mixed infections.

The bulk of the early work on the wild-primate malarias was done in the 1960s and 1970s by researchers including Robert G. Coatney, William E. Collins, Peter Contacos, Leonidas Deane, Don Eyles, P. C. C. Garnham, and McWilson Warren. In 1971, Coatney and associates published, *The Primate Malarias.* Although nearly 40 years have passed since its original publication, it remains the most comprehensive source of data on nonhuman primate malarias. Their typology of the primate plasmodia derived from morphological analysis and life history traits. Much of our knowledge about the specific phylogenetic relationships among the primate malarias has come to light only quite recently with advances in molecular genetics. Although this research has revolutionized our understanding of the relationships among these plasmodia, it is still a relatively new field, and many questions remain. Differing methods of analysis of genetic markers have sometimes produced differing phylogenetic trees of relatedness, particularly among the complicated relationships of members of the vivax group. Some relationships have been consistent across analyses, the closest relative of human *P. vivax* being, surprisingly, New World monkey *P. simium* (implicating a lateral transfer).

The molecular research that has been conducted on the relationships among the malaria plasmodia is biased toward falciparum and vivax types, primarily because these are the most significant in causing disease in human populations. Falciparum malaria, as stated, is the most severe form of malaria to humans and has been the best studied. Human vivax malaria, while being less severe, is the most frequently occurring type of malaria outside tropical Africa. Relatively less research has been done on the milder malariae types and ovale types, which appear to be less common worldwide. However, as is discussed later, the prevalence of *P. malariae* may be seriously underestimated in the New World tropics. Many of the wild-primate malarias remain molecularly uncharacterized, and their relationship to the others remains to be fully understood.

One of the most important early relationships that was identified was between two falciparum-type malarias: human *P. falciparum* and African ape *P. reichenowi*. Up until 2010, *P. falciparum* was thought to be a uniquely human parasite. However, as is discussed in Chapter 3, it has recently been discovered that bonobos (*Pan paniscus*), chimpanzees (*Pan troglodytes*), and gorillas (*Gorilla gorilla*) harbor *P. falciparum* and several additional falciparum-type plasmodia, previously unknown. Among the four types of malaria parasites in primates falciparum type is an outgroup (Chapter 3). The vivax, malariae, and ovale types are all more closely related to one another than any of the three are to the falciparum malaria.

Early hypotheses about the origin of falciparum malaria included it arising as a zoonosis from domesticated birds with the advent of agriculture. Some of the early genetic studies seemed to support this view. However, subsequent molecular studies demonstrated the close genetic relationship between *P. reichenowi* and *P. falciparum,* ruling out the avian hypothesis (McCutchan et al. 1984; Waters, Higgins, and McCutchan 1991, 1993). The prevailing view then became that the two forms of falciparum malaria separated with the divergence of the humans, chimpanzees, and gorillas 5 to 10 million years ago and that the two plasmodial forms co-evolved in humans and African apes.[6] In 2010 that view was upended several times by research suggesting that chimpanzees, bonobos, or gorillas were the original source of falciparum in humans (Chapter 3).

The relationships among the vivax-type malarias are extremely complicated and far from being definitively resolved for several reasons. One involves the state of development of genetic research. As previously stated, differing methods of genetic analysis have not produced identical phylogenetic trees. Although molecular studies have consistently placed the same plasmodia within the vivax group, there is controversy regarding the precise relationships among many of the plasmodia in the vivax group. Three factors make sorting out these relationships complicated: in some cases, the same primate host harbors multiple closely related vivax species, the same vivax species can infect multiple primate hosts, and vivax-type malarias are prone to host-switching. Thus co-evolutionary processes can be taking place in one plasmodia species across multiple hosts, and a single host can be involved with simultaneous co-adaptation to multiple plasmodia. And genetic relationships can be difficult to differentiate from convergent evolution when one is dealing with such closely related primate species and closely related plasmodia species. The possibility of potentially multiple lateral transfers during a given plasmodium's evolutionary history makes these relationships even more difficult to disentangle.

The vivax group includes at least 11 plasmodia, most of which occur in Southeast Asian primates. Included are seven closely related plasmodia of Asian monkeys, one occurring in gibbons (*Hylobates*). At least three forms exist outside Asia. One is the extremely widespread human vivax malaria that has historically affected humans outside tropical Africa in warm climates but that has also created infections as far north as Siberia. The distributions of the other two species are extremely puzzling. One is *P. simium,* which has been identified in three species of New World monkeys in a small geographic location on the Atlantic coastal forest of Brazil. As previously mentioned, molecular analyses have identified the closest relative of *P. simium* to be human *P. vivax*, which indicates a recent host-switch, since New World monkeys have been separated from other primate species for at least 30 million years (see, for example, Rosenberger, Hartwig, and Wolff 1991; Takai et al. 2000). Although it is clear that a lateral transfer has occurred, one study by Li and associates (2001) suggests that *P. simium* clusters with a variant of Old World *P. vivax* more closely than it does to the extant variant of *P. vivax* found in New World humans.

This situation raises the possibility that *P. vivax* was introduced twice into South America, with one variety becoming established in monkeys and another variety becoming established in humans. The issue will be discussed at length in Chapter 4, but suffice it to say here that such a finding poses a rather complicated problem to explain in terms of ethnological and archaeological evidence. Some have suggested that it indicates that there was a Pre-Columbian introduction of *P. vivax* malaria that became established only in a few monkey species; it would have had to have come from humans in Southeast Asia. In terms of the archaeological record, such a scenario is far from parsimonious and in fact seems rather far-fetched. Recent archaeological evidence suggests that Polynesians contacted South America in Pre-Columbian times, which involves traveling an incredible distance by boat. But southeastern human contact with South America would necessitate effectively doubling that distance, which seems quite unlikely, particularly given the absence of malaria in Fiji until recent times.

Equally puzzling is the existence of *P. gonderi*, a vivax-type malaria of mangabeys and the drill in Africa. Like New World *P. simium*, it exists in only a few species of monkeys in a relatively small geographic area on the Atlantic coastal forest of western Africa. Although it is likely significant that monkey *P. simium* and *P. gonderi* are appearing in biogeographically similar environments, given the oceanic divide and evolutionary divide, it is unlikely that there is a direct historical relationship between the two. The existence of *P. gonderi* in African monkeys is also puzzling, because apart from *P. vivax,* no other vivax-type malaria has been demonstrated in tropical Africa, with the possible exception of great ape *P. schwetzi,* which has been hypothesized to belong to either the ovale or vivax group. The widely distributed vervet monkey (*Cercopithecus aethiops*) and several species of baboons (*Papio* spp.) have been infected with *P. gonderi* under experimental conditions, but it seems to naturally occur only in a very restricted area in restricted species. The patterns of absence of malaria parasites are also significant, although not completely understood. For example, among Asian monkeys, macaques (*Macaca* spp.) appear to be far more heavily parasitized than Asian colobine monkeys. According to LeClerc and associates (2004), malaria is not known in the so-called odd-nosed colobine monkeys of Asia that are sympatric with macaques, including the Proboscis monkey (*Nasalis larvatus*), the Douc langur (*Pygathrix nemaeus*), and the simakobu (*Simias concolor*).

HOST-SWITCHING/LATERAL TRANSFERS

The evolutionary history of a host may not track with the evolutionary history of its parasite. Page and Charleston (1998) described for types of historical associations that can create complicated gene trees: co-divergence, duplication, horizontal transfer, and sorting. Co-divergence involves both the host and the parasite undergoing an evolutionary branching from a parent form to sister forms. Here, the host and the parasite undergo co-evolution, but the sister hosts and the sister parasites undergo independent evolution. In duplication, a parasite diverges into two or more sister forms within the same host. Horizontal transfer involves host-switching of a parasite into a new host. Sorting includes extinction, whereby a parasite form dies off. These processes are not mutually exclusive, because all can be characteristic of the history of a host or a parasite. The result is that although hosts and parasites may co-evolve over time, they can also undergo evolutionary processes that are independent of each other so that host trees and parasite trees may not match.

Garamszegi (2009) analyzed statistical patterns among the events due to co-divergence, duplication, host-switching, and extinction in primate malaria. He found that host-switches played a central role in shaping the distribution of malaria species across their primate hosts. However, it was also found that related lineages of primate plasmodia tend to infect hosts within the same taxonomic family. Thus, parasites do not host-switch randomly but tend to transfer to closely related species. Exceptions were noted for two species that have crossed great phylogenetic distance: *P. brasilianum* (a human to New World monkey lateral transfer) and for *P. knowlesi* (an Old World monkey to human lateral transfer). The implication is that although one can say that most plasmodia host-switches involve related species, flexibility exists in at least some species that makes the possibility of humans acquiring novel wild-primate malarias difficult to predict—or for that matter, for any one species of primate being infected with the plasmodia of any other primate species. Host-switching clearly can create thorny problems in reconstructing the evolutionary histories primate malarias.

The primate malaria parasites seem to display a good deal of variability in their host-specificity. For example, *P. brasilianum* has been documented in over 30 species of New World monkeys (see Appendix 1) and may very well be present in all New World monkeys. At the other extreme, *P. simiovale* has been documented only in the Toque macaque (*M. sinica*); *P. hyobati* has been documented only in the silvery gibbon (*Hylobates moloch*); and *P. pitheci* appears to infect only orangutans (*Pongo pygmaeus* s.l.). Such "host-specificity" may actually reflect geographic specificity; for example, New World *P. simium* infects only three primate species, all in the Atlantic forest. Moreover, it apparently infects only black howlers (*Alouatta carya*) living in the Atlantic forest, and it has not yet been found in populations of black howlers living in other parts of the Amazon. The distribution of vivax-type malaria in African monkeys and falciparum-type malaria in African great apes also seems to have geographic specificity, to be discussed further. Thus, historical ecological variables affecting the malaria landscape mush also be taken into account.

Among Asian primates where a vivax-type malaria has been identified, many have been demonstrated to serve as hosts for more than one species of plasmodia; for example, the crab-eating macaque, the Toque macaque, and the pig-tailed macaque (*M. nemestrina*) all serve as hosts to four different species of vivax plasmodia (but not identical subsets of the same four plasmodia). The white-handed gibbon (*H. lar*) is also infected by four species of malaria parasites. These have not been definitely classified as vivax type but are suspected to be of the vivax group, because only vivax-type parasites have been identified in Asian wild primates. Similarly, the orangutan (*Pongo pymaeus*) is infected with two species plasmodia (*P. pitheci* and *P. silvaticum*) that are also likely members of the vivax group.

Thus, in characterizing the primate malarias, many of the plasmodia species simultaneously evolve over multiple primate hosts. Some of the plasmodia have laterally transferred across distantly related species. It is also the case that some primate species harbor multiple plasmodial species, some of which are distantly related to one another. While Asian primates seem to harbor only vivax-type malarias, New World monkeys are naturally infected with malariae and vivax types, and African apes are naturally infected with falciparum, malariae, and ovale types (and possibly vivax). Malaria parasites are largely absent from Old World monkeys, with the exception of mangabeys and drills. However, many

Old World monkeys are parasitized by *Hepatocystis*, a sister taxa to the primate plasmodium species (see, for example, Martinsen, Perkins, and Schall 2008). Ecological factors are also at play in understanding the distribution of the primate malarias.

Although the primate malarias cannot be characterized as a whole as species-specific or even infraorder-specific for species that infect humans and New World monkeys, that does not mean that co-evolutionary processes are not at work in the relationship of a given individual primate species and an individual plasmodium species. In Chapter 2, genetic adaptations to malaria are discussed. The malaria hypothesis, proposed by J. B. S. Haldane in 1949, suggested that the thalassemias might bear a relationship to human resistance to falciparum malaria. Since then, a number of potential human genetic adaptations to malaria have been described, including sickle-cell trait and G6PD (Glucose-6-phosphate-dehydrogenase deficiency).

Among the nonhuman primates, to date we do not have clear evidence for genetic polymorphisms such as sickle-cell trait having arisen to provide resistance to malaria. However, that does not mean that adaptive mechanisms are not in place. Indirect evidence from studies involving experimental infections of various primates with novel malaria parasites are suggestive of co-evolutionary processes. For example, bonnet macaques (*M. radiata*) are natural hosts of *P. fragile* and experience only mild symptomology. However, Rhesus monkeys (*M. mulatta*) and Hanuman langurs (*Presbytis entellus*) become critically ill when experimentally infected.

Susceptibility and virulence are complicated among the primates and should be interpreted with caution (Chapters 2 and 3). In the case of *P. falciparum*, New World monkeys and African apes are susceptible, whereas Old World monkeys are not. Apparently an independent mutation arose in humans and New World monkeys that made both groups susceptible to *P. falciparum*. In the case of humans, the mutation seems specifically linked to co-evolutionary relationships with malaria parasites. However, that cannot possibly be the case for New World monkeys, because *P. falciparum* was introduced to South America only within the last 500 years with colonization, and New World primates are not hosts to any known falciparum-type malaria. Thus, the selective pressure that created susceptibility to *P. falciparum* in New World monkeys seems to be unrelated to *P. falciparum* itself. RBC Duffy negativity may be another example, which is widespread in tropical Africa and provides immunity to vivax-type malaria in humans. At this point, it is unclear if RBC Duffy negativity is a genetic adaptation to vivax malaria or if it arose under differing selective pressures and only consequentially provides the benefit of vivax resistance.

In addition, virulence itself is a very slippery slope for interpreting co-evolutionary processes. This is particularly the case for vector-borne diseases, whereby co-evolutionary processes occur not only across the host and pathogen but also with vector. Although many host-pathogen relationships are characterized by the tendency for hosts to become more resistant and pathogens to become more benign, this is not always the case. With vector-borne diseases, it is customary to identify three players: pathogen, host, and vector. Arguably, these are largely anthropocentric categories delineated from the perspective of human disease. From the plasmodium's point of view (so to speak), vector and host are not really meaningful distinctions. Plasmodia need both mosquito and vertebrate "hosts" to complete their life cycles. It might very well be more accurate for plasmodia to be considered more mosquito-adapted than vertebrate-adapted, because the most fragile part of the infectious chain lies with short-lived mosquitoes. From this perspective, it might be more logical to reverse the categories and deem humans as vectors and mosquitoes as hosts. Virulence in the vertebrate may even be adaptive if it benefits the viability of the mosquito-plasmodium relationship. In the case of *P. falciparum*, if human hosts are abundant, any specific infected individual, for all intents and purposes, need live only as long as a mosquito for a plasmodium to complete its life cycle.

The Ten-Thousand Year Fever

Malaria is a manifestation of complex co-evolutionary and biogeographical processes among vertebrates, mosquitoes, and plasmodia that have been at play for at least 30 million years. Nonetheless, my position is that a watershed human cultural event, the advent of agriculture, so dramatically reshaped human (and consequentially, wild primate) relationships with the malaria parasites and mosquito vectors that primate malaria is best conceptualized in the entirety of its evolutionary history according to two historical periods: pre- and post-circa 10,000 years ago. The advent of agriculture not only involved a change in the mode of human subsistence patterns but is also the starting point for a major shift in human relationships with the environment that have forever altered both human and wild-primate relationships with malaria parasites. Rather than humans moving nomadically to obtain resources, they began to become sedentary, domesticating plants and animals, altering the environment, and ultimately, changing the primate malarial landscape.

The archaeological evidence demonstrates that the advent of agriculture occurred in the context of climate change and is associated with human population expansion. One long-standing question in anthropology is if agriculture arose as a response to population expansion or if population expansion was a consequence of the development of agriculture. That is to say, did human groups begin to adopt agriculture because of population growth, requiring new means of subsistence, or was population growth a consequence of adopting a more productive means of subsistence?

Regardless of causation, human population growth and the beginnings of agriculture clearly are correlated. Agriculture (the domestication of plants and/or animals) arose independently in the Americas, Asia, and Africa. Thus climate change is strongly implicated in the shift in human cultures from hunting and gathering to agriculture, for varied plants and animals specific to local contexts have been involved. Thus, the advent of agriculture is not a solitary "invention" that can be traced to one specific human group that spread culturally to all human groups. Rather, beginning around 10,000 years ago, agricultural began to be adopted independently by many human cultures. Although the domestication of specific plants and animals did spread among geographically contiguous peoples, there are multiple worldwide loci for the adoption of agriculture.

Although it is conventional in anthropology to speak of the "agricultural revolution," this usage is to some degree a figure of speech. Using the benchmark of 10,000 years ago is meant to convey the approximate time when this process began. It is important to be clear that the shift of most human cultures from a hunting and gathering mode of subsistence to an agricultural mode of subsistence did not happen suddenly but was a gradual process. In addition, some human groups are long-standing hunter-gatherers, and others, such as the Guajá with whom I worked, reverted from agricultural to hunting and gathering. Nonetheless, the so-called "agricultural revolution" represents a starting point for the setting in motion of a fundamental shift in a great many of the world's cultures from adapting to the environment to meet their needs to humans adapting the environment itself to meet their needs. Within the last 500 years, this process has escalated. Colonization, development, and deforestation have so altered the primate malarial landscape that relationships that were previously evaluated in terms of evolutionary time must be reconceived in terms of the rapid changes in historic time.

NOTES

1. Notwithstanding the recent discovery of *P. knowlesi* host-switching, to be discussed later.
2. The nomenclature for classification of apes and humans is currently inconsistent in the literature. In the past, the term "hominid" was reserved for humans, human ancestors, and putative ancestors and their close relatives (such as the genus *Australopithecus*). Given the evidence from molecular biology of the close

relationship between humans and chimpanzees and humans and the African apes, conventional terminology is changing. In this text, "hominoid" refers to all apes and humans (including both the great apes [chimpanzees, gorillas, and orangutans] and the so-called lesser apes [gibbons and siamangs]); "hominid" refers to great apes and humans; "hominine" refers to humans and African great apes (chimpanzees and gorillas); "hominin" refers to human ancestors, putative ancestors, and close relatives of putative ancestors.

3. The order Primates (or Primata) is traditionally gradistically divided into Anthropoidea (humans, monkeys, and apes) and Prosmii (including lemurs, lorises, galagos, and tarsiers). A cladistic taxonomy that more accurately represents evolutionary relationships is the division between the Haplorrhini (humans, apes, monkeys, and tarsiers) and Strepsirrhinii (the prosimians except the tarsier). I am using the anthropoid/prosimian distinction, because this work focuses on humans, monkeys, and apes. Several forms of malaria-causing plasmodia have been identified in prosimians (see Appendix 1). However, virtually no information is available regarding the relationship of prosimian malaria to the other primate malarias. Therefore, the scope of this work is limited to "anthropoid" primates.
4. *P. schwetzi* has not been definitively classified but has been suggested to be a member of either the vivax or ovale types.
5. The exceptions are the owl monkeys (*Aotus* spp.), who are nocturnal (active at night) anthropoids, and the ring-tailed lemur (*Lemur catta*), who are diurnal (active during the day).
6. The suggestion of a division between human and African ape lines is problematic and represents a gradistic rather than a cladistic classification. Evidence from human evolution suggests that humans and chimpanzees share a more recent common ancestor than either does to gorillas. This issue is discussed at greater length in Chapter 3.

Chapter 2

Co-Evolution: Parasites, Vectors, and Hosts

Primates, mosquitoes, and plasmodia involve complex co-evolutionary relationships. Some have a deep evolutionary history, being at play long before the rise of the order Primates. Other relationships seem to have only recently emerged, perhaps within the last few thousand years, and likely are linked to human cultural changes since the advent of agriculture. This chapter begins with an overview of the life cycle of the plasmodium in primates and mosquitoes. Co-evolutionary relationships then are examined, with a focus first on plasmodia and mosquitoes. Next, the evidence for biological adaptation to malaria is discussed for both human and wild primates. The best-known adaptations are the human polymorphisms that violate Hardy-Weinberg equilibrium such as sickle-cell trait, RBC Duffy negativity, thalassemia, and G6PD.

The research to date suggests that nonhuman primate populations do not appear to exhibit polymorphisms analogous to those found in humans that are implicated in malaria resistance. That is by no means to suggest that wild-primate populations have not evolved any adaptive response to malaria. Rather, the evidence suggests a qualitative difference in the selective pressures in relation to malaria when one is comparing humans to other primate species. Given that human and wild primates are infected with closely related plasmodia, differences in their adaptive response lends weight to the position taken in this work that human cultural changes related to the rise of agriculture have dramatically altered human-plasmodia relationships. Moreover, human-mediated ecological changes also have consequences for the malaria landscape for all primates in creating conditions that may foster cross-infections among humans and wild primates, and among varied wild-primate species.

THE LIFE CYCLE OF THE PLASMODIUM

In the previous chapter, I argued that the convention of describing malaria as a disease involving a parasite, vector, and host is somewhat anthropocentric. A case can be made for describing the life cycle of malarial parasites as taking place within two hosts, a mosquito and a primate. Although discussion of the life cycle of plasmodia here refers specifically to primates, broadly, the process is generally the same in vertebrates. The basic stages in the life cycle of plasmodia have been described by numerous researchers (for example, Centers for Disease Control and Prevention 2006; Desowitz 1991; Humphreys 2001; Miller, Good, and Milon 1994; Sherman 2005; Sturm and Heussler 2007). A brief overview follows.

The plasmodium life cycle takes place in both a female *Anopheles* spp. mosquito and a primate. Since plasmodia require both organisms to reproduce, it is somewhat arbitrary where one begins the description of the cycle. Here, the starting point will be the "parents." A primate infected with malaria has the sexual forms of the parasite, the gametocytes, circulating in the blood stream. When a mosquito takes a blood meal from an infected primate, the gametocytes are ingested. Within the mosquito gut, sexual reproduction takes place resulting in sporozoite offspring. The sporozoites then migrate to the salivary gland of the mosquito.[1] Once the sporozoites are in the salivary glands, the parasite can be transferred to another primate when the mosquito takes another blood meal.

When the mosquito bites, her saliva makes contact with primate blood. Mosquito saliva contains vasodialating anticoagulant, and antihistamine compounds that facilitate feeding and maximize the quantity of blood that can be quickly ingested. Once in the bloodstream, the sporozoites migrate to the liver and infect hepatocytes. Here, they mature into schizonts, rupture, and release daughter cells (merozoites). The merozoites cover themselves in the cell membrane of infected liver cells to escape detection by the body's immune system and travel to the bloodstream. They are, in a sense, behaving like wolves wrapped in sheeps' clothing (merozoites wrapped in liver clothing), because the body does not recognize the merozoites as pathogenic when they use a liver cell membrane cloak.[2]

The merozoites develop in the bloodstream in two ways. Some merozoites mature into gametocytes. Mosquitoes who bite when circulating merozoites are available in a primate's bloodstream allow for the start of a new life cycle of the plasmodia within the mosquito. Other merozoites asexually reproduce and create more merozoites. The asexually reproduction of merozoites involves invasion of the red blood cells, which is responsible for the disease symptoms of malaria. The merozoites break out of the liver in periodic waves, which coincide with the cyclic fevers characteristic of malaria.

Two mechanisms are involved in the disease state, release of toxins and damage to the erythrocytes. When the red blood cells rupture in periodic waves, toxins are released into the body that are largely responsible for the associated cyclic fever and chills. In addition, both direct and indirect consequences of damage to red blood cells contribute to the disease state of malaria. The human circulatory system can be thought of as something akin to a highway system. It functions as a route of transport for the raw materials that are needed for the body to function. Major blood vessels, such as the aorta, are the main highways, and the capillaries are like the neighborhood side streets. One of the main materials that the body needs to have transported to its cells is oxygen. The red blood cells are the vehicles that transport the oxygen. Essentially, these broken-down red blood cells are like stalled vehicles all over the highway. The problem is not only a lack of a sufficient quantity of oxygen-carrying vehicles but also the problems caused by damaged red blood cells that clog the road and make it more difficult for the healthy red blood cells to move freely. The initial consequences are anemia and hypoglycemia. In severe malaria, it can lead to renal failure, pulmonary edema, cerebral anoxia (in cerebral malaria), and death.

One important variation on the life cycle pattern is the existence of a dormant stage. In some forms of malaria, a dormant form of the parasite exists between the sporozoite and the merozoite stages, called a hypnozoite. Dormant stages are found in *P. vivax* and *P. ovale* in humans, and in *P. cynomolgi, P. simiovale*, and *P. fieldi* in Asian monkey malarias (Coatney et al. 1971; Cogswell 2000). As is discussed in more detail later, the hypnozoite stage seems typically to involve winter hibernation in temperate climates, with malaria relapsing when the weather warms and mosquitoes are actively biting. The upward limit of how long a hypnozoite can remain dormant is not clear. And the trigger for hypnozoites may not be exogenous climate change but mosquito-mediated. In other words, in most populations with seasonal changes in mosquito densities, it might appear that malaria relapses seasonally. The actual key variable may not be so much climatic, but hypnozoites may have the ability to remain dormant indefinitely and be "awakened" only by the stimulus of biting mosquitoes.

From the Plasmodium's Point of View

Regarding co-evolutionary processes involving malaria parasites, greatest attention is usually given to the potential human adaptations that have evolved to confer some degree of resistance to the disease of malaria. However, co-evolutionary processes have to be understood as occurring across primates, vectors, and plasmodia. A variety of selective pressures and adaptations exist that primarily involve the mosquito-plasmodium interaction. On one hand, mosquitoes can be considered a far more tenuous part of the infection chain, because they are short-lived, and a much narrower window of opportunity exists for plasmodia to complete the mosquito stage of the life cycle. On the other hand, what mosquitoes lack in the limitation of the life span of any given individual, they more than make up for in terms of their voluminous raw numbers of individuals.

Some evidence exists that mosquitoes are preferentially attracted to individuals who are already infected with malaria. This suggests that coevolutionary processes are at play, whereby plasmodia are able to alter the behavior of mosquitoes. In a study on Kenyan children, those harboring *P. falciparum* gametocytes attracted twice as many mosquitoes as those who were either uninfected or were infected but not actively harboring the gametocyte stage of the parasite (Lacroix et al. 2005). Here, it appears that the parasite is able to drive mosquitoes to bite at a time most favorable for transmission from a human to the mosquito, thereby increasing the potential for plasmodial reproductive success.

The presence of the malaria symptoms may be beneficial for plasmodia reproductive success. Nacher (2005) has described the fever, anemia, and fatigue in human hosts as parasite "manipulations" to lure mosquitoes and satisfy plasmodia needs. Among humans, feverish hosts display a number of characteristics that attract mosquitoes such as rapid breathing (releasing more CO_2) and exposure of more skin surface area (when clothing and covers are minimized to dissipate heat). Another factor that may be attractive for the mosquito is that the feverish host is sweating and may have higher levels of kairomones (compounds that attract mosquitoes, discussed in more detail shortly). But perhaps the most important factor in terms of immediate selective pressure on mosquitoes is that feverish individuals are sluggish and less likely to effectively swat and kill mosquitoes.

The manipulations of the plasmodia present an interesting set of coadaptations that may be competing with one another across players. The fever response to disease has often been interpreted as beneficial owing to its ability to stimulate the immune system and, for some diseases, to create a temperature level inhospitable to pathogens (Blatteis 2003). There is some evidence that the fever response may be advantageous in humans for combating malaria infection. For human *P. falciparum*, at least in vitro, the parasite appears to be inhibited by febrile temperatures (Long et al. 2001). Although fever is a consequence of red blood cell destruction, it also creates hosts that are more attractive to mosquitoes, not only biologically but also behaviorally, because hosts that are ill and fatigued provide less dicey targets. But as Ewald (1994) has described, the fact that fever may be beneficial against one kind of pathogen does not mean that it is beneficial to all pathogens; shutting down the fever response for one type of pathogen might also shut it down for all. His implication is that the fever response is a kind of statistical game theory. If fever is generally beneficial in fighting off pathogens, there is selective pressure for the fever response, even if it is detrimental for a minority of pathogens.

Nacher (2005) has also argued that the timing of the fever in malaria is synchronized with the stage in the life cycle of the plasmodia that is most beneficial for successful transmission. The fever occurs when merozoites burst from red blood cells, but this is also the time when gametocytes are present in the blood. According to Nacher, given that the mosquito has to survive at least 13 days for the sporozoites to mature, the timing is quite critical. Fever induces mosquitoes to bite at the right time. Thus, plasmodia survival depends on the inducing the mosquito to bite during a critical window of opportunity, and so both vector and host variables come into play (the "extended phenotype," see Dawkins 1999).

To put it a little more loosely, the plasmodia need not be concerned with inducing mosquitoes to bite in general. A female mosquito is genetically prewired and driven to take blood meals, because her eggs cannot develop without blood proteins. From maturity until death, adult females repeat a two- to three-day cycle of taking a blood meal, resting, laying eggs, and then taking a blood meal again. What is critical for the survival of the plasmodia is that a female mosquito bite a host with gametocytes in his or her bloodstream *early* in her life. If that does not occur, then there may not be sufficient time for the plasmodium to complete the portion its life cycle that must take place in the mosquito. A mosquito may live only a few weeks, and it takes nearly that long for the gametocytes to mature in her body.

Evidence exists that plasmodia affect the feeding behavior of mosquitoes depending on the developmental cycle of the plasmodia within the mosquito. In a study on the feeding persistence of *Anopheles stephensi* infected with *P. yoellii nigeriensis*, it was found that mosquitoes increased attempts to feed on hosts when sporozoites were present in the salivary glands but decreased feeding behavior when they were harboring oocysts, meaning that sporozoites are still developing (Anderson, Koella, and Hurd 1999). Since sporozoites are transmissible and oocysts are not, it appears that the plasmodia are modifying the feeding behavior of the mosquitoes to increase the likelihood of biting when the plasmodia are in a stage that would allow them to be transferred to the host. This is potentially detrimental to the mosquito if feeding behavior is being controlled to some degree by the needs of the plasmodia rather than the needs of the mosquito.

Although it is also necessary for the mosquito to bite twice to transfer the matured gametocytes (sporozoites) into a human or other host, that does not present as dire a problem. It is a given that the female mosquito will be taking blood meals late in life and for as long as she is alive. Since with the second bite she is now dropping off plasmodia rather than picking them up, so to speak, the time factor is not as critical, because primate hosts (even with malaria) live longer than mosquitoes.

This raises the issue of virulence. For many infectious diseases (but not all), co-evolutionary processes over time may lead to pathogens becoming less virulent and for hosts becoming more resistant to pathogens. To paint this scenario with a very broad brush, in some host-pathogen dyads, an equilibrium evolves so that host survival in a population is enhanced by increased resistance to the pathogen, and pathogen survival is enhanced by decreased virulence, so that the disease state does not kill the host. For example, during the era of European colonization, contracting measles might be a rather unpleasant experience for European populations long adapted to this disease, but it was life-threatening when introduced to indigenous New World peoples with no previous exposure, where such equilibrium had not been reached.

Although the study of virulence is actually more complicated that the simple scenario described above, when studying vector-borne diseases, we have to consider all conditions and potential co-adaptations of the triad: vector, host, and pathogen. Although virulence might be detrimental to a primate host, it may be beneficial to the parasite. Alternatively, in some contexts, virulence might very well be irrelevant to plasmodia survival. As a given, plasmodia must be mosquito-adapted, and mosquitoes live only a few weeks. Again, the plasmodia may need for the host to survive only about as long as a mosquito in order to effectively reproduce. This depends on the demographic and ecological context. For human falciparum malaria, a largely (but not always) tropical disease whereby mosquitoes thrive year round, host virulence may not be an issue for plasmodia reproductive success. For human vivax, which can survive in colder climates, decreased virulence may be beneficial so that the plasmodia can survive in dormant form in the host until ecological conditions are present to favor effective transmission.

McNeill (1976), who is renowned for his work on describing how Old World diseases decimated New World peoples, also hypothesized early on that vector-borne diseases presented far more complicated conditions. He argued that "benignness" could be accomplished in only the vector or the host, but not both. His reasoning was that since short-lived vectors are more critical in the triad, the tendency was for benignness to evolve in vector-parasite relationships, with human "hosts" being more peripheral to the equation.

Ewald (1994) operationalized this concept as the *restricted adaptation hypothesis*. He proposed an alternative hypothesis to explain virulence called the *adaptive severity hypothesis*. He argued that vector-borne diseases tend to be more virulent than non-vector-borne diseases, because this situation benefits the vector. One of the major differences he notes between vector- and non-vector-borne diseases is that in the case of the latter, the host needs to be well enough to move and spread the disease to others. That is not necessary in vector-borne diseases, because immobilization of the host does not impede transmission, because the mosquito does the job of spreading the disease from person to person. Moreover, he points out that virulent vector-borne pathogens tend to reach higher densities than less virulent pathogens do, which means that mosquitoes can be more readily infected. Therefore, he essentially argues that in vector-borne diseases, virulence in the "host" may actually be beneficial to the parasite.

As is discussed in more detail later, the virulence of human *P. falciparum* in comparison to *P. vivax* malaria has led some to suggest that it provides evidence for *P. falciparum* being a more recent human parasite than *P. vivax*. Given that virulence can be adaptive to the parasite in diseases involving the parasite/vector/host triad, high levels of virulence do not necessarily correlate with time depth of evolutionary exposure to a pathogen. In other words, although the documented pathogenicity of measles to New World peoples can be explained through recent host-pathogen relationships, malaria is not a comparable analogy, because we must consider all potential permutations of vector, host, and pathogen relations.

If pathogenicity is not a reliable indicator of age, then we must still ask why malaria is life threatening in some cases and benign in others. We cannot simply make a comparison between human falciparum and human vivax, because human vivax can also present as a quite virulent disease. One possible explanation is that adaptive strategies, for host/vector/parasite, are largely dependent on the specific ecological context. As previously described, for some of the malarias, such as human *P. vivax* and *P. ovale*, and several of the vivax-type wild-primate malarias, the potential exists for the parasite to enter a hypnozoite stage. The possibility should be considered that virulence and hypnozoite potential are functionally equivalent plasmodial adaptations. Ultimately, both serve to effectively synchronize the timing of mosquito biting for the benefit of plasmodia survival. However, that still does not explain why human *P. malariae* and *P. ovale*, and many of the other wild-primate malarias, are relatively benign.

Ewald (1994) offered the intriguing possibility that in some vector-borne diseases, host virulence may be the reflection of long-term adaptation of the parasite rather than benignness. One example he uses is the increasing severity of Asian monkey *P. knowlesi* when cycled through several generations of human hosts (Chapter 6). If virulence can indicate *older* age of host-pathogen relationships, then it could suggest that more benign forms of malaria are younger. However, the situation does not appear to be so straightforward. Reactions from experimental primate hosts infected with novel forms of malaria are all over the map with benign infections in some and lethal infections in others (Chapter 6). Chapter 3 discusses recent findings that suggest that the virulence of human *P. falciparum* is a consequence of a previous adaptation that conferred resistance to *P. reichenowi*. The co-evolutionary relationships involved in malaria are complicated and still not completely understood, with multiple factors at play at the level of host, mosquito, and plasmodia that lead to highly variable adaptations in different species and different environmental contexts.

From the Mosquito's Point of View

Mosquitoes are generally described as vectors, or "carriers," of malaria, because the infection with the parasites in mosquitoes does not produce an analogous disease state to that found in primate hosts. However, one cannot conclude that the relationship of the plasmodia to the mosquitoes is entirely benign. Studies investigating the biological effects of malaria parasites on mosquitoes have produced mixed results: some studies have not demonstrated significant effects on mosquitoes, whereas others have demonstrated decreased survival rates.

The strict laboratory controls often used to study levels of virulence in mosquitoes may actually be factoring out the critical variables that determine differences in effects among mosquito populations. Ferguson and Read (2002) have argued that one reason for the mixed results in the literature on the effects of malaria parasites on mosquitoes is that insufficient attention has been given to genetic variability among mosquitoes and to variability in environmental conditions that may affect the level of virulence in the mosquito vector. In a study on the effects of rodent *P. chaubaudi* on the vector *An. stephensi*, it was found that vector mortality varied significantly between two mosquito genotypes and that virulence increased when glucose water was limited (differences were also found in the level of virulence between the two genotypes when glucose water was limited). Glucose water is equivalent to the availability of flower nectar, required by both male and blood-feeding female mosquitoes.

Two main effects were found in a study of *P. falciparum* sporozoite infection in *An. gambiae* mosquitoes (Anderson, Knols, and Koella 2000; Rossignol, Ribeiro, and Spielman 1984; Wekesa, Copeland, and Mwangi

1992). First, plasmodia infection impaired salivary function, making it more difficult for mosquitoes to find blood vessels. As a consequence, female mosquitoes probe more often. Second, infected mosquitoes were more likely than non-infected mosquitoes to bite more than once and to return after being shooed away. Both effects create increased mortality risk during feeding. Thus, infection increased high-risk behavior in mosquitoes in that they spent more time probing and bit more often. While such risk-taking behaviors on the part of the mosquitoes could serve to enhance parasite transmission, they also diminish survival for any given mosquito.

In considering potential mosquito adaptations one must consider that although malaria is global, mosquitoes are local. Mosquitoes seldom venture more than a mile from where they hatch. They are short-lived, but because they reproduce in such vast numbers, they have proven to be highly adaptable to any given local conditions. Although some generalizations can be made about anophelines, the behavior of a given species can vary greatly depending on the particular set of environmental circumstances. However, it should be stressed that research on malaria and anopheline mosquitoes is for the most part geared to understanding the relationship to human hosts. Very little is known about the vectors of wild-primate malaria. Most of what is known about wild primates and their vectors comes from research conducted in the 1960s and 1970s. However, with the recent *Plasmodium knowlesi* outbreak, and the recent identification of *P. falciparum* and new falciparum-type malaria species in African apes, I suspect that in the next few years, funding will be available to support research in this area.

Nonetheless, one generalization that can be made at this point is that anopheline mosquitoes do not bite completely at random. Even highly anthropophilic mosquitoes demonstrate preferences for some humans over others. Although such behavior suggests that some selective force is driving them to make "choices" among their potential options, the basis for such "choices" is not well understood. One ecological model used to describe mosquito preferences is ideal free distribution (IDF) (Fretwell and Lucas 1969). As applied to mosquito behavior (Kelly 2001), if a mosquito has equal access to host X and host Y, and there are no differences between the two, each would have an equal chance of being bitten. If the bite rates differ, then selective pressure is operating, because the mosquito would not waste time bypassing one equally attractive host to get to another. IDF is somewhat similar to optimal foraging theory (MacArthur and Pianka 1966), which looks at variables such as range, food selection, and time spent feeding in order to predict behavior in terms of costs and benefits. But when one is describing female mosquitoes seeking a blood meal, IDF can be conceptualized according to the simpler dynamic of choosing a host.

Numerous studies have demonstrated that anopheline mosquitoes display preferences based on such variables as age, sex, ABO blood type, pregnancy, and the disease state of the host. Again, although anophelines have been consistently shown to demonstrate preferences, they are highly variable even within a particular species in a local context. Host differentiation can be as broad as being anthropophilic (preferring humans) or zoophilic (preferring animals) to demonstrating a clear preference for one individual over another when no known discernible differences can be identified.

A number of studies have been conducted that describe anthropophilic versus zoophilic preferences of mosquitoes. For example, one of the most anthropophilic mosquitoes is *An. gambiae*, the major vector of falciparum malaria in Africa (see, for example, Coluzzi et al. 1979; Coluzzi, Petrarca, and di Deco 1985). In contrast, in the lowlands of Bolivia, *An. pseudopunctipennis* feeds opportunistically on either humans or animals but prefers humans, sheep, goats, and donkeys—but dogs, pigs, and chickens are rarely bitten (Lardeux et al. 2007). The behavior of any particular species of mosquito can also vary significantly locally. For example, in the context of rice cultivation in Kenya, *An. arabiensis, An. coustani, An. pharoensis*, and *An. funestus* all showed a high preference for bovines over humans (Muriu et al. 2008). Few studies have been done on mosquito preferences and behavior among nonhuman primates; most have been conducted in experimental settings.

Regarding human populations, it has long been known that many anthropophilic anophelines, such as *An. gambiae* and *An. albimanus*, demonstrate a preference for adults over children and adult males over adult females (Muirhead-Thomson 1951). Some anopheline mosquitoes also display a preference for pregnant women over nonpregnant women. For example, studies with *An. gambiae* in Gambia and *A. arabiensis* in Sudan demonstrated that pregnant women attracted twice as many anopheline mosquitoes as nonpregnant women (Himeidan, Elbashir, and Adam 2004; Lindsay et al. 2000). For both of these studies, no significant differences were found in the biting rates of non-anopheline mosquitoes.

One possible explanation is that pregnant women are simply larger than nonpregnant women, providing more surface area. But Lindsay and associates (2000) suspect that CO_2 production may be more important than host size and weight. In Gambia, they found that women 28 weeks into their pregnancies produced 21% more exhaled air than their nonpregnant counterparts. Although CO_2 production is a reliable indicator that breathing hosts are near, it may not play as big a role in attracting mosquitoes as once thought. For example, in a study comparing *An. gambiae* attraction to human breath and CO_2 alone at similar concentrations, it was found that mosquitoes demonstrated only slight preferences for human breath (de Jong and Knols 1995). One study suggested that some individuals may have components in their breath that act as a mosquito *repellent* for *An. gambiae,* diverting the mosquitoes toward other parts of the body (Mukabana et al. 2004).

What may be playing a bigger role are human kairomones. Kairomones are analogous to pheromones, but while pheromones attract desirable others (such as mates or pollinators), kairomones attract unwanted predators. The human kairomones that attract mosquitoes are probably the product of odors and compounds resulting from the interaction of gland secretions and skin microorganisms that can vary from individual to individual (Braks, Anderson, and Knols 1999). Specifically, it appears that *An. gambiae* is very much attracted to bacterial compounds produced by human foot odor as well as limburger cheese, which produces similar compounds (de Jong and Knols 1995). Thus CO_2 or other human breath volatiles may initially attract *An. gambiae* but also repel mosquitoes away from the head, so that they then drawn to the feet, where they are less likely to be effectively swatted. Kairomones may also be involved in preferences that relate to hormonal differences that affect glandular secretions beginning at puberty that may play a role in the preference of adults over children and males over females (Braks, Anderson, and Knols 1999). Such hormonal differences could play a role in attraction to pregnant women.

Anthropophilic anophelines also display a preference for certain individuals over others, which remains poorly understood. Several studies have been done using a mosquito trapping device called an olfactory tent and humans protected in nets as bait. In two similar studies done on Kenyan men with *An. gambiae* (Mukabana et al. 2002) and Tanzanian men with *An. gambiae* and *An. funestus* (Knols, de Jong, and Takken 1995), significant differences were found for attraction to individuals. Both these studies involved men of the same age in the same village with no discernible relevant differences among them in terms of diet, life history, or other factors. If given equal access and equal choice, the mosquitoes demonstrated definite preferences for certain individuals over others. The basis for such mosquito choices is not known but may involve differences in individual kairomone signatures.

Human Adaptations to Malaria

In 1949 Haldane proposed that the high frequencies of the blood disorder thalassemia in malarious areas were related to resistance to malaria. This proposal has become known as the "malaria hypothesis," and since then a number of other polymorphisms have been described that are implicated in co-evolutionary processes associated with malaria. These genetic variations relate to the Hardy-Weinberg principle and balanced polymorphism (Solomon, Berg, and Martin 2008). Hardy-Weinberg equilibrium describes an ideal state in large populations where genotypes and allele frequencies remain stable over time. Violations to Hardy-Weinberg equilibrium can occur when there is nonrandom mating, in or out group migrations within a population, or selective pressure. In the case of the suspected adaptations to malaria, unusually high frequencies of certain alleles and genotypes occur in a population, suggesting selective pressure. Many of the adaptations involving balanced polymorphism occur where heterozygotes for an allele have an advantage, which leads to higher frequencies that expected of uncommon alleles. The heterozygous state may also involve a dominant normal allele and a partially expressed abnormal recessive allele. One consequence of selection of the heterozygote is that some percentage of the population will also inherit two abnormal alleles, leading to genetic disorders such as sickle-cell anemia. In truth, historically it has been the presence of anomalous high frequencies of homozygous genetic disorders that has often led to discovery of heterozygote adaptive benefits.

Genetic variations that have been linked to malaria resistance include sickle-cell trait, α and β-thalassemia, glucose-6-phosphate-dehydrogenase deficiency (G6PD), RBC Duffy negativity, and hereditary ovalocytosis. Although there is evidence that these polymorphisms provide resistance to certain types of malaria and that they exist at higher frequencies in malarious areas than in nonmalarious areas, it does not necessarily prove that the polymorphisms were positively selected for in response to malaria. The alternative hypothesis is that these polymorphisms are the result selection pressures unrelated to malaria. The majority of the adaptive benefits are for virulent *P. falciparum*. The one notable exception is RBC Duffy negativity, which as will be discussed, confers complete resistance to *P. vivax* malaria. It also differs from other polymorphisms in that it is a dominant homozygous state that has reached near fixation (universality) in tropical Africa.

Several of the polymorphisms involve hemoglobin. The two major categories of inherited hemoglobin disorders involved are structural variants that produce abnormal hemoglobins and the thalassemias that create an imbalance in the ratio between α and β hemoglobins. In terms of structural differences, over 700 structural variants of hemoglobin have been identified; of these only three (Hb S, Hb C, and Hb E) occur in high frequencies in human populations (Weatherall and Clegg 2001).

Sickle-cell trait is perhaps the best known of the human adaptations that provide resistance to malaria. The sickle-cell gene is widely distributed in sub-Saharan Africa, the Middle East, and in parts of India (Weatherall and Clegg 2001). Sickle-cell trait was first hypothesized to have a role in malarial resistance by Allison (1954). Allison described the relatively high frequencies of sickle-cell trait among many populations in tropical Africa and the Mediterranean and questioned why it would be maintained at such high rates in these areas. He estimated that the random mutation rate for sickle-cell trait would need to be on the order of 3,000 times what it would be expected to be in order to account for the number of people with the trait. Since carriers of the trait (the heterozygous state) did not seem to suffer any ill health effect, he hypothesized that a selective advantage could be operating among carriers that perpetuated the mutation. He did a small study of "sicklers" versus "non-sicklers" and found significantly lower rates of falciparum malaria among the sicklers. It has since been confirmed that heterozygotes have a 10-fold reduced risk of falciparum malaria (Kwiatkowski 2005).

Sickle-cell anemia occurs when an individual possesses two abnormal Hb S genes, which results in deformed (sickle-shaped) red blood cells that can form dangerous blockages in the capillaries (Solomon, Berg, and Martin 2008). Heterozygotes, under conditions of normal living, do not generally suffer the ill health effects experienced by homozygotes but still exhibit a degree of sickling. Individuals with sickle-cell anemia, without medical intervention, usually die before they reach reproductive age. Evolutionary parsimony and logic should result in natural selection operating to decrease the frequency of sickle-cell trait rather than to raise it. One explanation for how resistance is achieved in the carrier is that parasitized blood cells seem to be more likely to sickle in carriers, which causes them to be removed from the blood stream at an higher rate than that of nonparasitized cells (Luzzatto, Nwachiku-Jarrett, and Reddy 1970). Thus, for carriers, sickle-cell trait does not prevent them contracting falciparum malaria but makes its effects less severe. Today, the trait is maintained at about 10% frequencies in some regions where malaria is endemic (Kwiatkowski 2005).

Two less common forms of abnormal hemoglobin are Hb C and Hb E. Higher-than-expected frequencies of Hb C occur in West Africa and Hb E in Southeast Asia and southern India (Kwiatkowski 2005; Weatherall and Clegg 2001). Hemoglobin C has been demonstrated to provide protection against falciparum malaria for both the homozygous and the heterozygous states (Modiano et al. 2001). The homozygous condition does not appear to manifest in any clinical disease symptoms and confers stronger protection against falciparum malaria than does the heterozygous state. At this time, it is not completely clear whether Hemoglobin E confers resistance to malaria (Weatherall 2008). However, computer simulations suggest that in Thailand, the Hb E variant derives from a single origin, 1,240–4,440 years ago, and that the mutation spread rapidly, implying that it was under strong selective pressure (Ohashi et al. 2004). Malaria is certainly suspect as a selective agent, but this has not yet been proven.

The thalassemias are a group of inherited genetic disorders that involve defective synthesis of either the alpha (α thalassemia) or beta (β thalassemia) chains of hemoglobin (Weatherall 2008). It is found in the Mediterranean, Africa, the Middle East, southern India, Southeast Asia, Melanesia, and the Pacific Islands (Weatherall and Clegg 2001). Although sickle cell involves a mutation that leads to the production of an abnormal type of hemoglobin, the thalassemias involve an imbalance in the ratios of the alpha and beta hemoglobin chains produced (see Solomon, Berg, and Martin 2008). Hemoglobin consists of iron-porphyrin (or heme) bound to a protein called a globin. Most humans produce roughly equal amounts of two types of globin, alpha globin and beta globin. The structure of hemoglobin typically consists of two alpha globin chains and two beta[3] globin chains attached to a heme. In alpha thalassemia, the body produces too few alpha globins; in beta thalassemia, the body produces too few beta globins. The consequence in both is abnormal oxygen transport by the red blood cells and increased vulnerability of the red blood cells to injury and destruction. Both types result in anemia.

The mechanism that provides resistance to malaria in thalassemia carriers is not well understood. Paradoxically, children with alpha thalassemia on the Pacific Island of Vanuatu display increased susceptibility to vivax malaria, but vivax infection appears to act as a natural vaccine against falciparum malaria (Williams et al. 1996). Another way that thalassemia may provide protection against malaria is through alterations in red blood cell size and raw numbers. Fowkes and associates (2008) found that children in Papua New Guinea with alpha thalassemia have unusually small red blood cells, but produce 10–20% more red blood cells. They note that a severe attack of malaria can result in the loss of one-third to one-half of the total red blood cell count. These individuals' red blood cells contain less hemoglobin, but they have a larger number of red blood cells overall. Thus, those with mild alpha thalassemia suffer less severe results from malaria anemia, because they end up with more overall hemoglobin after an attack than those who do not have the genetic condition.

Other genetic variants that confer resistance to malaria are red blood cell membrane variations; RBC (red blood cell) Duffy negativity is the best known. RBC Duffy negativity is a homozygous state of red blood cell antigens that confers complete resistance to *P. vivax* but not to *P. falciparum* (Marsh 2002). *P. vivax* requires what is termed the Duffy antigen receptor for chemokines (DARC) to be present on the surface of a red blood cell in order to be able to invade (Hamblin and Di Rienzo

2000). Over 97% of the indigenous populations in sub-Saharan and West Africa exhibit RBC Duffy negativity (Carter and Mendis 2002), with *P. vivax* infection occurring at a rate of less than 1% (Mendis et al. 2001). Those cases of *P. vivax* that do occur in this area have been judged to be largely attributable to either immigration into the region or from indigenous inhabitants returning from areas where *P. vivax* is common. RBC Duffy negativity does not occur in high rates in populations outside Africa, except for areas with large numbers of African-descended immigrants (Cavasini et al. 2007; Hamblin and Di Rienzo 2000).

Because RBC Duffy negativity is virtually universal in tropical Africa and confers complete resistance to *P. vivax*, some researchers believe this provides compelling evidence that human vivax malaria is very ancient disease that originated in Africa. Here, RBC Duffy negativity is viewed as a human adaptive response that eventually eliminated susceptibility to *P. vivax* in tropical Africa. This view is bolstered by the discovery of a variant form of Duffy negativity in Papua New Guinea (Zimmerman et al. 1999). However, it is not yet clear whether or not this variant of Duffy negativity offers any significant protection against vivax malaria. One way that the Papua New Guinea form differs from the African form is that it has been identified only in the heterozygous state, although it occurs more typically in the homozygous state in Africa. Although Papuan New Guinea Duffy negativity appears to be an independent parallel evolution, and malaria resistance is implicated, research to date has not determined that it provides an adaptive benefit.

Despite the fact that RBC Duffy negativity in Africa confers complete resistance to vivax malaria, its evolutionary relationship to malaria remains debated. Livingstone (1984) offered two hypotheses. One is that RBC Duffy negativity was selected for because it provided resistance to vivax malaria. The other is that RBC Duffy negativity was selected for owing to a reason unrelated to vivax malaria, but its presence prevented vivax malaria from becoming established in tropical Africa. He favored the latter hypothesis, because he was skeptical that strong enough selective pressures would have existed for the comparatively more typically benign form of vivax, unless it had been more severe in the past. It should also be considered that because Duffy negativity is nearly universal in Africa but rare in non-African populations, it might have arisen within the last 100,000 years, after the migration of modern humans out of Africa. However, Zimmerman and associates (1999) comment that in light of the Papua New Guinea form of Duffy negativity, vivax malaria might be under stronger selective pressure than previously realized. They make the case for this situation based not only on the fact that RBC Duffy negativity arose independently in Papua New Guinea but also that it could act to undermine mechanisms that provide resistance to falciparum malaria, since thalassemia appears to decrease the offer resistance to falciparum by increasing susceptibility to vivax.

Two less widely distributed forms of red blood cell membrane variation that likely provide protection against falciparum malaria are Gerbich negativity and hereditary ovalocytosis. Individuals with the Gerbich negativity mutation lack glycophorin C, which is a protein of the red blood cell membrane that acts as a receptor for *P. falciparum* (Maier et al. 2002). Gerbich negativity occurs in approximately 45% of the populations of coastal Papua New Guinea, where malaria is endemic (Weatherall 2008). Another Melanesian polymorphism is ovalocytosis, which involves a structural variant in a red blood cell protein (Kwiatkowski 2005). Melanesian ovalocytosis occurs in up to 10% of the population of Papua New Guinea; it is autosomal dominant and appears to be lethal in homozygotes (Allen et al. 1999; Genton et al. 1995). Heterozygotes are actually more prone to attacks from falciparum malaria, but they are almost completely protected against cerebral malaria (Allen et al. 1999).

Glucose-6-phosphate-dehydrogenase deficiency is an enzyme disorder that is particularly important in the metabolism of red blood cells (Capellini and Fiorelli 2008). The most serious symptom is hemolytic anemia, which is most commonly triggered by eating fava beans, but drugs or infection can also initiate the reaction (Capellini and Fiorelli 2008, Luzzatto 2006). It is an X-linked disorder, so males are hemizygous, whereas females may be homozygous (normal or deficient) or heterozygous (Luzzatto 2006). In both hemizygous males and heterozygous females, individuals with falciparum malaria have lower parasite loads than do those without the mutation (Allison and Clyde 1961; Ruwende and Hill 1998; Ruwende et al. 2002). G6PD is common in areas where falciparum malaria is endemic, with the highest prevalence rates occurring in tropical Africa, the Middle East, tropical and subtropical Asia, and some areas of the Mediterranean and Papua New Guinea (Capellini and Fiorelli 2008, Nkhoma et al. 2009, Weatherall and Clegg 2001). In Kenya, New Guinea, Greece, Sardinia, and Papua New Guinea, prevalence rates are higher at altitudes below 1,000 meters, which are more malarious (Ruwende and Hill 1998).

Another genetic variable that may relate to malaria resistance involves the ABO blood groups. The geographic distribution of the ABO blood groups has long been suspected to have a relationship to malaria (Athreya and Coriell 1967; Cavalli-Sforza and Feldman 2003). Several studies have demonstrated that in the Old World, type O occurs at higher frequencies near the equator and that type A occurs at higher frequencies in colder zones; in addition, the course of falciparum malaria is less severe in type O individuals (see, for example, Cserti and Dzik 2007, Fischer and Boone

1998; Lell et al. 1999). Cserti and Dzik (2007) hypothesize that this may be due to reduced cytoadhesion of red blood cells in type O individuals. The adhesion of infected red blood cells within small blood vessels is a property of *P. falciparum* that can have particularly severe consequences in pregnancy, when such sequestration of infected red blood cells can occur in the placenta (Steketee et al. 2001). In studies of pregnant women in Gambia and Malawi, type O blood was associated with an increased prevalence of placental malaria in primaparous women (first pregnancy), but multiparous women had a reduced risk (Adam et al. 2007; Loscertales and Brabin 2006; Senga et al. 2007).

Wild-Primate Adaptation to Malaria

Given the multiple types of genetic variations that have been identified in humans that offer resistance to malaria, we might expect that similar adaptations would have arisen in wild primates infected by related species of malaria parasites; however, this does not appear to be the case. Very little evidence exists that suggests that any similar polymorphism can be found in wild primates. A major obstacle to effectively assessing potential polymorphisms is the dearth of epidemiological data on malaria in primate populations. The primary reason why such evidence is lacking is logistical. The type of surveillance programs that are done in human populations are extremely difficult to duplicate in wild primates. Wild primates must be first located, trapped, and sedated in order to collect the needed blood samples. Although a few excellent studies have been done, unfortunately most have had very small sample sizes, which make extrapolation to the general population difficult.

This situation may very well soon change. As described in Chapter 3, only within the last year of the writing of this work, advances in molecular genetics have produced technology to identify malaria infection from primate fecal samples. It is truly a watershed change that has potential to revolutionize our understanding of the primate malarias. Collecting the regularly produced fecal material from wild primates is relatively cheap and noninvasive and avoids the need to trap and sedate wild primates to collect blood. Although the technology has not yet been used to identify potential polymorphisms, that day is soon coming, and we will then have a much better understanding of whether or not wild primates have genetic adaptations to malaria.

One of the few studies that have been done involved determining if a G6PD-type polymorphism existed in chimpanzees (Verrelli et al. 2006). Using a sample of 46 unrelated chimpanzees, Verrelli and associates (2006) investigated whether they exhibited any evidence of a similar enzyme mutation. No evidence was found to suggest a similar G6PD mutation in chimpanzees. They interpreted the results to indicate that the G6PD polymorphism in humans is both recent and unique. In a similar study involving three subspecies of the chimpanzee, no evidence could be found of deleterious mutations in (β-globin) or G6PD that would suggest that they have similar adaptive mechanisms as found in human hemoglobin variants or in the G6PD enzyme (MacFie et al. 2009).

In a study of orangutans (Steiper et al. 2005), polymorphisms were found in α-2 globin that deviated from Hardy-Weinberg equilibrium. The researchers suggest that it may represent an adaptation to malaria. However, although such deviation from Hardy-Weinberg suggests selective pressures, what cannot be determined at this point is whether the selective pressure is due to malaria or some other factor. It is also worth recalling here that Williams and associates (1996) found that alpha thalassemia in Vanuatu actually *increased* susceptibility to vivax-type malaria while decreasing susceptibility to the more lethal falciparum-type malaria. As previously mentioned, although the two species of orangutan malaria parasites have not been definitively classified, it is strongly suspected that they belong to the vivax group. (Steiper and associates classify *P. silvaticum* and *P. pitheci* as vivax-type plasmodia.) In addition, it should be noted that the Bornean orangutan is endangered, and the Sumatran orangutan is critically endangered, thus the small population size itself could be a factor in assessing deviations from Hardy-Weinberg equilibrium. Nonetheless, it is an intriguing study, and far too little research is available of this type.

One reason why understanding possible genetic adaptations of wild primates is important is that there is potential to provide insight into human relationships. If the relationship of humans to malarial plasmodia changed with the advent of agriculture and changing environmental conditions, then wild primates can potentially provide an analogy for pre-agricultural human populations. At this point, we do not have evidence that wild primates exhibit polymorphisms similar to that found in humans that provide resistance to malaria.

Several possible explanations exist for this situation. One possibility is that insufficient data is available to fully assess whether wild primates have such polymorphisms. They may exist, but they have not yet been documented. Another possibility is that similar polymorphisms are not observed in wild primates because there is no analog to the virulence of *P. falciparum* in humans. Verrelli and associates (2006) make the point that wild-primate malarias are more similar to human ovale, malariae, and vivax in that the symptoms are relatively benign in comparison to human falciparum malaria. Even the long-recognized falciparum-type malaria in African apes, *P. reichenowi*, does not exhibit severe symptoms in the African great apes (Rich et al. 2009). For the most part, the polymorphisms seen in humans confer resistance to falciparum malaria, the key exception being RBC Duffy negativity and *P. vivax*. Thus, wild-primate malarias may not be virulent enough to lead to such polymorphisms. Several researchers have noted that the lethality of *P. falciparum*, particularly with children before reproductive age, allows for the rapid emergence of polymorphisms (Cserti and Dzik 2007; Kwiatkowski 2005).

A final possible explanation is the recent expansion of malaria in human populations. Although not mutually exclusive with other explanations, it could provide a working hypothesis to explain the differences in genetic adaptations in humans that do not yet appear to have an analogue in any wild primate. Again, although human and wild primates are infected with related plasmodia, they do not seem to exhibit similar adaptive responses that provide resistance. That does not mean that wild primates do not possess biological adaptations to malaria. In Chapter 6, experimental studies with wild primates are discussed that do seem to suggest differences in specific primate species susceptibility and resistance to specific plasmodial species. Rather, wild primates do not appear to have evolved the same type of genetic adaptations against malaria. Although malaria is ancient in primates, the evidence points to a recent change in the relationship of humans with malaria parasites, particularly falciparum malaria. It is likely linked with cultural changes associated with the rise of agriculture, but such human-mediated environmental changes have also affected wild-primate populations in multiple ways, to be discussed further in this work.

Notes

1. It has recently been discovered that transmembrane protein MAEBL is needed for *P. falciparum* to be able to invade a mosquito's salivary glands (Saenz et al. 2008). Genetically modified mosquitoes that had the MAEBL ligand deleted were unable to invade the salivary glands.
2. See Sturm et al. 2006 and Sturm and Heussler 2007 for a description of the process.
3. Broadly, the protein portion of the hemoglobin molecule is typically composed of alpha and beta globins. However, humans also produce a smaller percentage of gamma and delta globins that can substitute for beta globins. For this reason, the hemoglobin molecule is often described as formed by the combination of an alpha chain and a non-alpha chain (mainly beta, but also gamma and delta) that pair with each other at about a 1:1 ratio to form normal hemoglobin.

Chapter 3

Falciparum Type: The Great Ape Malaria

The best studied of the human malarias is the deadliest form, *P. falciparum*. According to the World Health Organization (2009), a child dies of malaria every 30 seconds; 247 million malaria cases occurred in 2006, causing nearly 1 million deaths, mainly among African children. Most of the deaths are due to *P. falciparum*. Falciparum malaria differentially affects children. In untreated falciparum malaria, children have 20 times the case fatality rate as adults (Cserti and Dzik 2007). In 2002 it was estimated that there were half a billion clinical episodes of falciparum malaria (Snow et al. 2005).

Falciparum-type malaria is designated here as the ancestral hominid malaria, because its only known living natural hosts are humans, bonobos, chimpanzees, and gorillas. However, both the origin and the age of falciparum-type malaria have been the subject of much debate. Various hypotheses have put the time of entry of falciparum malaria into human populations at dates ranging from 5,000–5 million years ago. Such wide variations in hypotheses of origin in humans are largely due to the limitations of the methods available to establish evolutionary relationships among malaria-causing plasmodia. The earliest hypotheses posited that falciparum was laterally acquired from birds and was associated with the introduction of agriculture within the last 10,000 years. Some of the first genetic studies suggested that falciparum malaria diverged at the time when the human line diverged from the *Pan* (chimpanzee and bonobo) and *Gorilla* lines, approximately 5 million years ago. The codivergence hypothesis has held sway until only the last several years, when new evidence emerged that has suggested a relatively recent host-switch of falciparum-type malaria from one of the great ape species to humans. In addition, several lines of evidence are converging to suggest that regardless of the ultimate origin of falciparum-type malaria in humans, it underwent a population expansion in humans, becoming a serious epidemic disease. Although this chapter seeks to evaluate the available evidence to date on the origin of falciparum malaria in humans, which is still much debated, *The Ten-Thousand Year Fever* posits that malaria may have only recently become a serious epidemic in humans.

Our current understanding of the relationships among all the plasmodia that cause primate malaria (human or wild) rests on the advances in molecular genetics. The bulk of this research has taken place only within the last 10 years. Before the rise of this discipline, researchers relied on gross comparative anatomy of plasmodial species or on life-history characteristics. Both have proven to be misleading, particularly life-history characteristics that include features such as periodicity and hypnozoite stages. As previously described, periodicity refers to characteristic waxing and waning of the fever cycle in malaria. The human malarias are characterized by cycles of fever and fever remission, which occur at different intervals, depending on the species. Early malariologists thought that species with similar cycles shared a recent common ancestor, but apparently this is not the case. Periodicity is plastic; species that share a common ancestor may not have the same number of days in a cycle, and more distantly related species may have similar patterns of periodicity. Similarly, the presence of hypnozoite stages in some species were at one time considered to represent shared evolutionary history. In some forms of malaria (but not present falciparum type), a dormant liver stage exists within the human/wild primate side of the life cycle of the plasmodia, whereby the organism goes into stasis. This, too, has been demonstrated to be plastic; it is not necessarily a result of shared evolutionary history but often parallel evolution, with the same strategy apparently having evolved independently in several plasmodial species.

In the review of the molecular genetics for falciparum malaria in this chapter and vivax in the next chapter (very little research has been done on *P. malariae* or *P. ovale*), keep in mind that molecular genetics is a research area that is still in its infancy. Even within the narrow time frame of my research on this book (over the past six years), assumptions have been repeatedly upended with new findings. This chapter provides a description of the advances and limitations of methods used in phylogenetic reconstruction, including the current state of mapping of the *P. falciparum* genome.

THE AVIAN HYPOTHESIS

One of the earliest hypotheses that argued for a recent origin of falciparum malaria is "avian hypothesis." It is suggested that falciparum malaria was first introduced to human populations by way of lateral transfer from birds. Morphological similarities in gametocytes of avian and human falciparum malaria had been observed as early as 1890

(McCutchan et al. 1996). Initially, these similarities between human and avian forms were interpreted in two ways: either that *P. falciparum* was the oldest of the human malarial parasites and ancestral to *P. vivax, P. malariae,* and *P. ovale* or that the human malarias are polyphyletic[1] (Sinden et al. 1978). In other words, on the one hand, all human malarias may have originated from avian malarias, with *P. falciparum* being the earliest and the other human malarias later deriving from human falciparum (or as a group are sister taxa). On the other hand, human falciparum and the other human malarias are only distantly related to one another.

As previously described, researchers relied on comparative anatomy and life-history characteristics to classify the malaria parasites before the emergence of molecular genetics (see Coatney et al. 1971). The life-history approach involves evaluating the behavior of the parasite in the host and includes features such as periodicity, hyponozoite stages, and virulence. All can be misleading, particular life-history features, which can be quite variable and may be explained more by parallel evolution rather than common origin. McCutchan and associates (1984) was one of the first studies to challenge the significance of morphological and life-history features for classification of the malaria parasites. It had long been assumed that human *P. falciparum* and two Asian monkey parasites, *P. fragile* and *P. coatneyi,* were all closely related to one another (Coatney et al. 1971). McCutchan and associates (1984) compared the nucleotide sequences of several *Plasmodium* genomes and found that *P. fragile* was actually more closely related to Asian monkey *P. knowlesi* (in the vivax group) than to *P. falciparum*. Their findings link *P. fragile* to vivax-type malarias and have been confirmed by multiple later genetic studies.

However, McCutchan and associates (1984) also found that *P. falciparum* appeared to be more closely related to the avian/rodent malaria clade than to the other human and monkey malaria parasites they compared. At this point, they interpreted the finding cautiously. Although their findings seemed to suggest that a close relationship between *P. falciparum* and the avian parasites could be due to host-switching, they also suggested an alternative hypothesis that *P. falciparum* was quite ancient and shared primitive features with the avian parasites. At the very least, the findings suggested significant genetic distance between human falciparum and members of the vivax group. The study was significant, because it was among the earliest to provide a genetic basis for understanding the relationships among the primate malaria parasites and also a potential way to determine the time depth, because it was one of the first genetic studies to suggest a close relationship between avian plasmodia and human falciparum.

Another study in the early history of molecular phylogenetics of plasmodia compared primate parasites using the small subunit (SSU) rRNA genes[2] (Waters, Higgins, and McCutchan 1991, 1993). Their findings suggested that human falciparum and the avian plasmodia were monophyletic and formed a separate clade from that of the other primate parasites (as well as rodent parasites). They concluded that avian malaria parasites and human falciparum shared a common ancestor and that human falciparum derived from birds. They also linked their findings to the earlier anthropological work of Livingstone, who proposed a link between human falciparum and agriculture, and suggested that the domestication of birds in Africa provided an ecological context for close association between humans and birds that facilitated a lateral transfer.

Livingstone has often been cited in the literature as associated with the avian hypothesis (for example, Ayala et al. 1998; Escalante, Barrio, and Ayala 1995; Schlagenhauf 2004; Waters, Higgins, and McCutchan 1991, 1993; Wiersch, Maier, and Kampen 2005), referring to his influential article on the relationship of sickle-cell gene distribution and malaria (Livingstone 1958). However, he was not actually a proponent of the avian hypothesis. Rather, Livingstone is notable for proposing that *P. falciparum* became hyperendemic in tropical Africa owing to the advent of plant domestication. He argued that forest clearing for yam and rice cultivation provided ideal breeding grounds for anopheline vectors to thrive. Here, cultural behaviors were suggested to have had a profound influence on the population dynamics of mosquitoes, which created conditions that led to the subsequent proliferation of malaria in human populations. His argument was essentially epidemiological in attempting to explain how falciparum malaria became so widespread in tropical Africa.

Because Livingstone advanced an agricultural hypothesis, over time it has been erroneously conflated with the avian hypothesis. The critical dynamic for Livingstone was landscape change associated with the rise of agriculture owing to plant domestication, which led to changes in mosquito ecology. In the avian hypothesis, the critical dynamic is at the level of shared hosts: human and bird interaction. However, Livingstone did offer the possibility that plant domestication may very well have affected plasmodial host interactions, but not with birds. He suggested that disruption of the natural vegetation through agriculture could lead to the displacement and/or extinction of *mammals* and that their parasites might find a new host in humans. He later presciently proposed that human falciparum malaria was likely a lateral transfer from chimpanzees[3] (Livingstone 1971). Recent evidence generally supports his early hypothesis that humans first acquired falciparum-type malaria from an African ape host-switch, although the responsible species remains unclear.

The Cospeciation Hypothesis

The avian hypothesis was soon challenged by a subsequent study that also compared SSU rRNA genes but included African ape *P. reichenowi* in the analysis (Escalante and Ayala 1994).

The results indicated that *P. falciparum* and *P. reichenowi* were very closely related sister taxa and formed a separate clade from that of other primate parasites. The avian parasites were found to be only remotely related to *P. falciparum* and *P. reichenowi*. The SSU rRNA studies also proposed a divergence time for *P. reichenowi* and *P. falciparum* somewhere between 6–10 million years ago, consistent with the general time frame of the divergence of humans, *Pan* (chimpanzees and bonobos), and gorillas. They suggested that the common ancestor of falciparum-type plasmodia underwent cospeciation, with *P. falciparum* evolving in the ancestral human line and *P. reichenowi* evolving separately in the *Pan* and *Gorilla* lines. One problem with such a scenario, which will be discussed further, is superimposing behavioral "cladistics" with genetic "gradistics." Although contemporary *Gorilla* and *Pan* species are arguably behaviorally more similar to each other today than they are to contemporary humans (*Homo*), genetically the evidence suggests that ancestral humans and ancestral chimpanzees and bonobos share a more recent common ancestor than either group does to ancestral gorillas.

Qari and associates (1996) also analyzed primate, avian, and reptile plasmodia using the SSU rRNA genes and found similar results to Escalante and Ayala (1994). They proposed two major clades. One group consisted of bird and reptile plasmodia, and the second included all the primate malarias and rodent malaria. A key significance of these two molecular studies is that both seemed to rule out any possibility of the avian hypothesis that *P. falciparum* originated in humans from a lateral transfer from birds, as proposed by Waters and colleagues (1991, 1993). However, as will be discussed shortly, it now appears that the proponents of the avian hypothesis may have been correct in terms of the time frame for epidemic malaria in humans, correct in the mechanism of falciparum originating as lateral host transfer, but wrong in identifying the originating host species.

Perhaps the biggest question at this point is in attempting to understand how it could be that molecular genetic studies with SSU rRNA could lead to such dramatic differences in time frame? We have studies that point to an origin of human falciparum around 5–10 *thousand* years ago and others that are pointing to an origin closer to 5–10 *million* years ago. Two issues are relevant here that can help to understand such wildly divergent dates from genetic methods. One involves the nature of SSU rRNA in plasmodia and the other is a phenomenon known as "long branch attraction." First, as background, the small subunit ribosomal RNA is a molecule that exists in all known organisms and has been used to establish molecular phylogenies for a number of species (see review article by Solomon, Berg, and Martin 2008). However, plasmodia and related parasites are unusual in that they have multiple copies of the SSU rRNA gene that become active during different stages of the life cycle and do not evolve together (Corredor and Enea 1993; Perkins and Schall 2002; Roy and Irimia 2008). Thus, despite the recognized broad-based general value of the SSU rRNA gene for constructing molecular phylogenies, it appears to be a very poor locus for understanding the relationships among the malaria parasites.

The second issue that can help explain how the SSU rRNA studies provided remarkably different dates of divergence involves long branch attraction, which is particularly a problem due to the "A-T richness" (DNA bases of adenine and thymine) of plasmodia (Bergsten 2005; Telford and Copley 2005; Zilversmit and Hartl 2005). In long branch attraction, species that have been evolving faster than others tend to group together in phylogenies, regardless of their true relationship. It is essentially a problem of mistaking parallel evolution for common ancestry in closely related species. In the case of *P. falciparum*, long branch attraction resulted in its being erroneously linked as a close relative of the avian malarial parasites. It is complicated by the "A-T richness" problem. To begin with, the number of nucleotides is limited to 4 (and to 20 for amino acids). The theoretical algorithm of 4 factorial in calculating potential combinations of bases in practice for plasmodia genetics is based toward 2 factorial owing to A-T relative overrepresentation among the nucleotides. The simple solution to long branch attraction is to add more taxa. In the case of understanding the evolutionary position of human falciparum, adding the closely related *P. reichenowi* to the mix upended the genetic relationships, resulting in a radically different phylogeny.

Molecular studies using the circumsporozoite protein gene were next used to construct phylogenies of the primate plasmodia. Using this locus, Escalante, Barrio, and Ayala (1995) found confirmation of the close falciparum-reichenowi relationship and estimated a time frame of divergence of approximately 9 million years ago (consistent with the Escalante and Ayala [1994] time from of 6–10 million years ago). However, an avian circumsporozoite gene was not available to include in the analysis. McCutchan and associates (1996) did a similar study but included avian *P. gallinaceum* (found in domesticated poultry). They found support for Escalante group's separation of falciparum/reichenowi into a separate clade from the other primate malarias but also found that falciparum/reichenowi clade was more closely related to avian malaria parasites than it was to other human parasites. The study did not indicate that falciparum arose from gallinaceum but rather that they share a common ancestor. They proposed a three-clade model: falciparum/reichenowi and avian; rodent; and the other primate malarias.

The circumsporozoite protein has also been called into question as a suitable locus for establishing relationships among primate malaria plasmodia, because it is directly involved in the host immune response, and therefore may be under differential selective pressure (Zilversmit and Hartl 2005). Additional molecular studies have involved other loci. To address some of the difficulties with SSU rRNA, Escalante and associates (1997) conducted an analysis of SSU rRNA that incorporated secondary structure information. Their data suggested that the falciparum-reichenowi clade shares a common ancestor with avian-lizard malarial parasite. This lent support to McCutchan and associates' 1996 circumsporozoite gene studies. Cytochrome b has also been used as a locus. Escalante and colleagues (1998) found falciparum/reichenowi to be a sister to a cluster of *Plasmodium* and *Haemoproteus* parasites that infect birds. They also raised the possibility that *P. falciparum* arose from a host-switch from *P. reichenowi*. In a more comprehensive study, Perkins and Schall (2002) analyzed the relationships among 52 plasmodia using cytochrome b. Their research suggested that the immediate ancestor of *P. falciparum* was present at the time of the separation of the lineages leading to humans and chimpanzees. However, they found that falciparum-type malaria was not closely related to that of birds and reptiles. A study by Roy and Irimia (2008) using mitochondrial rRNA genes and rare genomic changes (loss/gain of splicesomal introns, mitochondria genomic indels, and nuclear gene loss) showed the primate malaria parasites to be linked to other mammalian malaria parasites but did not demonstrate a close relationship to the avian parasites. Martinsen, Perkins, and Schall (2008) also found primate malaria parasites as belonging to the same clade as that of the mammalian malaria parasites.

The emerging view of the molecular studies as of 2008 confirmed a very close relationship between *P. falciparum* and *P. reichenowi*. Although early studies provided conflicting information regarding the relationship of falciparum-type parasites to those of birds, reptiles, and other parasites, the preponderance of evidence suggested that there was not a close relationship between falciparum-type malaria parasites and avian parasites. Regarding falciparum-type malaria, the growing consensus was that *P. falciparum* and *P. reichenowi* diverged within the time frame of 5–10 million years ago, around the time of the divergence of humans and African apes, and thus the two plasmodial species evolved independently in these two host evolutionary lines.

An essential flaw in the cospeciation hypothesis was a failure to recognize anthropological evidence regarding the evolutionary relationships among human, *Pan* (chimpanzee and bonobo) and gorilla ancestors. Again, the distinction between "apes" and "humans" is gradistic rather than cladistic, representing western linguistic categories rather than actual genetic relationships among humans and the so-designated African great apes. In anthropology, the preponderance of evidence from molecular research in primate evolution for the last 40 years has pointed to *Pan* (chimpanzee and bonobo) and human ancestors as sharing a more recent common ancestor than either has with gorillas (for example, Caccone and Powell 1989; Chen and Li 2001; Chimpanzee Sequencing and Analysis Consortium 2005; Horai et al. 1992; Patterson et al. 2006; Ruvolo 1997; Wilson and Sarich 1969).

Thus, from genetic evidence, there is actually no such thing as a "human" line and an "ape" line that could accommodate a cospeciation hypothesis whereby falciparum and reichenowi diverged with "humans" and "African apes." Since *P. reichenowi* was described early on as occurring in both *Gorilla* and *Pan* species (Bray 1963; Coatney et al. 1971; Warren and Wharton 1963), it would seem to rule out the possibility of independent evolution of *P. falciparum* and *P. reichenowi* in human and "African ape" lines. If falciparum-type parasite divergence were to be coincident with the divergence of humans, bonobos, chimpanzees, and gorillas, we would not expect a division between human parasites and African ape parasites. Rather, we would have one parasite line in gorillas, who first diverged from the triad, and another parasite line common to *Pan* (chimpanzees/bonobos) and humans. That is not the case. Unless *P. reichenowi* was recently and independently acquired by chimpanzees/bonobos and gorillas, it would have been present in ancestral humans at least until the time of the divergence of human and chimpanzee/bonobo ancestors. Yet it seems that the variety of *P. falciparum* found in humans seems to be unique to our species. The cospeciation hypothesis can neither, on genetic grounds, account for why it emerged as an epidemic disease in our species nor can it yet provide a reliable time frame for when human falciparum first evolved in either contemporary humans or their ancestors. "Malaria's Eve" and "Big Bang" hypotheses offer explanations of how, when, and why falciparum malaria emerged in humans, which again point to the cultural variable of the agricultural revolution as key.

Malaria's Eve, Host-Switching, and Virulence

"Malaria's Eve" is a hypothesis that suggests that all extant human falciparum derived from a single, recent common ancestor. Owing to low levels of neutral polymorphisms, it was argued that a bottleneck effect had occurred in *P. falciparum* s.l., giving rise to contemporary *P. falciparum* s.s., within at least the last 200,000 years and perhaps within the last few thousand years (Mu et al. 2002; Rich et al. 1998; Volkman et al. 2001; see also Clark 2002; Su, Mu, and Joy 2003). Neutral polymorphisms or "silent mutations" are those that are presumed to be random mutations that are not undergoing selective pressure. Therefore, their rate of change is more likely to be constant over time. Low levels of such silent mutations suggest a relatively shallow time depth for falciparum malaria.

Malaria's Eve was actually not a direct challenge to the view that *P. falciparum* and *P. reichenowi* had diverged with humans and great apes. Rather, it suggested that regardless of the ultimate age of *P. falciparum*, one variety of falciparum in the human population replaced all others. Although the date of origin of Malaria's Eve has been debated, some researchers suggested that it arose within the last 10,000 years and was associated with human population expansion associated with the rise of agriculture (Rich et al. 1998; Volkman et al. 2001). Malaria's Eve was bolstered by evidence of a parallel event occurring in mosquito ecology and evolution. The major vector of human falciparum, *Anopheles gambiae*, also apparently underwent a recent speciation and expansion event in tropical Africa (Coluzzi et al. 1979; Coluzzi, Petrarca, and di Deco 1985). The *An. gambiae* expansion is thought to be due to landscape changes, altering the forest to agricultural zones. Broadly, Malaria's Eve suggested a big bang in the evolutionary history of human falciparum. Human cultural behaviors associated with the beginnings of agriculture were implicated in both changes in *P. falciparum* and *An. gambiae* natural selection and evolution.

Again, although Malaria's Eve suggested that contemporary *P. falciparum* arose from a recent common ancestor, it did not necessarily discount the hypothesis of ancient cospeciation of *P. falciparum* and *P. reichenowi* with the divergence of humans, gorillas, and chimpanzees/bonobos. Rather, it could encompass both an ancient cospeciation event and a more recent *P. falciparum* s.s. bottleneck and expansion event. Only within the last months of the first draft of this book has evidence emerged that all *P. falciparum* may have arisen in humans much more recently, by way of a lateral transfer from African apes.

Chimpanzees

Our understanding of the relationships among the falciparum-type malaria parasites (as well as all malaria parasites) is limited by the availability of malaria parasites that have been isolated, sequenced, and available for analysis. Before 2009 only one single isolate of *P. reichenowi* was known. Recently, Ollomo and associates (2009) collected blood samples from 17 chimpanzees in Gabon and found in two of them, a falciparum-like plasmodium. Analysis of its complete mitochondrial genome (including Cyt b, Cox I, and Cox III genes) indicated that the falciparum-like plasmodium was closely related to, but older than, the clade that includes *P. reichenowi* and *P. falciparum*. They named this recently identified "falciparum-like" potential new species in chimpanzees as *P. gaboni* sp. nov. (see also Duval et al. 2010; Prugnolle et al. 2010).

In a study published two months later (Rich et al. 2009), six additional isolates of chimpanzee falciparum-like plasmodia from Cameroon and the Ivory Coast were analyzed along with the "*P. gaboni*" identified by Ollomo and colleagues (2009). They found that *P. reichenowi* is more genetically diverse than originally known. Moreover, *P. falciparum* was described as fully included within the broader genetic diversity of *P. reichenowi*. They suggest that *P. falciparum* arose as a lateral transfer from chimpanzees, which could have possibly occurred as early as 2–3 million years ago, but may very well have occurred within the last 10,000 years.

An additional implication of the work of Rich and associates (2009) is that it has potential to shed light on the long-standing issue of understanding the significance of the virulence of *P. falciparum*. Historically, much of the debate surrounding the age and the origin of human falciparum has involved comparisons to human vivax in terms of both degree of virulence and potential co-evolutionary genetic adaptations that provide resistance to malaria. A number of early malariologists believed that human falciparum was of relatively recent origin owing to its high levels of pathogenicity in comparison to other human malaria parasites (for example, Boyd 1949; Coatney 1971). Bruce-Chwatt (1965) similarly made this argument but also offered the alternative hypothesis that sickle-cell trait could indicate a long-term adaptation. Livingstone (1958, 1971) viewed sickle-cell trait as a recent adaptation and very early on hypothesized a recent lateral transfer of *P. falciparum* from chimpanzees. The avian hypothesis (Waters, Higgins and McCutchan 1991, 1993) also suggested that the virulence of *P. falciparum* could be the result of a recent host-switch, but in their scenario, from birds. Ewald's (1994) adaptive severity hypothesis suggests that virulence in vector-borne diseases benefits the parasite and may reflect long-term adaptation in the parasite-host-vector triad.

Human vivax malaria has also been an important component of the debate. As described in Chapter 2, RBC Duffy negativity confers complete immunity to *P. vivax* malaria, but genetic adaptations to *P. falciparum* can be said to provide only increased resistance. In some cases, genetic adaptations to *P. falciparum* seem to operate, paradoxically, by increasing susceptibility to *P. vivax*. The effectiveness of RBC Duffy negativity in providing complete immunity to vivax malaria is arguably the result of long-term adaptation. However, it is also possible that RBC Duffy negativity arose under selective pressures unrelated to malaria. We should also take into account the negative case of New World monkey susceptibility to human falciparum malaria being due to an enzyme mutation (see following discussion) very similar to a mutation that makes humans susceptible to *P. falciparum* (Martin et al. 2005). Given that there is no evidence to suggest any possible exposure of New World monkeys to falciparum-type malaria before human colonization of the Americas within the last 500 years, the selection of the mutation that makes New World monkeys susceptible to falciparum malaria at this point seems to be unrelated to malaria itself. This fact provides another unresolved piece of the puzzle, because we do not know what selective pressures, ostensibly unrelated to malaria, gave rise to a mutation that made New World monkeys, like humans, susceptible to virulent falciparum infection. Ultimately, in humans, susceptibility to the virulent *P. falciparum* may be able to be explained according to the "Red Queen hypothesis."

In Lewis Carroll's *Alice's Adventures in Wonderland* (1865), an often quoted line by the Red Queen states, "it takes all the running you can do to keep in the same place." Although Rich and associates' (2009) study suggested that human falciparum originated in a host-switch from great ape *P. reichenowi*, the virulence of human falciparum does not appear to be simply the result of a recent lateral transfer. Rather, it seems that a mutation arose in human ancestors that provided resistance to *P. reichenowi*, and a subsequent mutation arose in *P. reichenowi*, that took advantage of the mechanism that had originally provided resistance in humans, leading to virulent *P. falciparum*. Some time after the separation of the human, chimpanzee/bonobo, and gorilla lines, an enzyme mutation arose in humans that conferred resistance to *P. reichenowi*. All contemporary humans possess this enzyme mutation. Subsequently, a mutation arose in *P. reichenowi*, representing the species *P. falciparum*, that allowed it to infect humans with this enzyme mutation more efficiently.

Human and chimpanzee DNA sequences are at least 96% identical, and perhaps as great 99% identical to one another (Chimpanzee Sequencing and Analysis Consortium 2005). For anthropologists, understanding the nature and significance of these differences is of particular interest in providing insight into human and primate evolution. In 1998 the first genetic difference that had a clear biochemical consequence in the evolutionary history of humans and the great apes was found in an enzyme mutation in humans involving sialic acids (Chou et al. 1998;

Irie et al. 1998; Muchmore, Diaz, and Varki 1998). Cells of extant free-living organisms are covered with a dense coating of glycan chains (sugars), forming a "glycocalyx" that Varki (2001), in his review article, has colorfully compared to the icing on a birthday cake. A family of sugars called sialic acids tends to be the outermost units on the glycan chains in vertebrates and some higher invertebrates. In mammals, two major forms of sialic acids are found: N-acetylneuraminic acid (Neu5Ac) and N-glycolylneuraminic acid (Neu5Gc). They differ in that Neu5Gc has an additional oxygen atom. These sialic acids are activated by conversion into sugar nucleotides. Neu5Ac is converted into CMP-Neu5Ac (cytidine monophosphate-Neu5Ac). Neu5Gc is converted to CMP-Neu5Gc by the action of an enzyme called CMAH (CMP-Neu5Ac hydroxylase). Humans have a mutation in the CMAH gene which led to a loss of Neu5Gc, creating Neu5Ac predominance.

The human CMAH gene mutation that led to Neu5Gc loss was suspected to be a random mutation that was selected for, because it provided resistance to pathogens that required cell-surface Neu5Gc to gain entry into the cell (Varki 2001). The hypothesis seems to be consistent with *P. falciparum*, because it differs from *P. reichenowi* in having a mutation that allows it to take advantage of the human Neu5Ac predominance. An additional factor in this relationship is that Martin and associates (2005) demonstrated that the major merozoite binding protein, EBA-175 (erythrocyte binding antigen-175), plays a central role in red blood cell invasion in most falciparum strains. They found that *P. falciparum* EBA-175 prefers Neu5Ac, whereas *P. reichenowi* EBA-175 prefers Neu5Gc; moreover, Neu5Gc interferes with *P. falciparum* EBA-175 binding. They also found a similar pattern of Neu5Ac predominance in *Aotus nancymaae*, suggesting parallel evolution of a similar mutation in New World monkeys.

These findings both shed light and raise additional questions about a number of patterns that are seen in *P. falciparum* and *P. reichenowi* in humans, great apes, Old World monkeys, and New World monkeys. It appears that the mutation that prevents expression of Neu5Gc in humans provides effective resistance to *P. reichenowi*. Although the evidence is scant, several early attempts that were made to infect humans with *P. reichenowi* were unsuccessful. The discoverers of *P. reichenowi*, Bray and Adler, inoculated themselves with the parasite in 1922 but were not able to establish an infection (Cohen 2009). In 1939 Rodhain attempted to infect two patients with *P. reichenowi* but was unsuccessful (Coatney et al. 1971). In addition, studies attempting to infect the vector *An. gambiae* with *P. reichenowi* have not been successful (Bray 1957 [cited in Coatney et al. 1971]; Collins et al. 1986). Similarly, it seems difficult to establish an infection of *P. falciparum* in chimpanzees. In one study, an infection was established in a splenectomized chimpanzee (Bray 1958a). In another, a *P. falciparum* infection was established in the chimpanzee only after the parasite had been modified (Taylor et al. 1985). In addition, in a six-month study of 68 chimpanzees and gorillas in a colony located in a region of Gabon with high rates of human *P. falciparum* infection, none of the great apes was ever found to harbor the parasite (Ollomo et al. 1997). However, as discussed below, *P. falciparum* has been recently identified in chimpanzees (Duval et al. 2010) as well as bonobos and gorillas (discussed shortly).

The timing of the host-switch still remains at issue. Although Rich and colleagues (2009) reject the cospeciation hypothesis of human and great ape falciparum-type malarias, which would place the date of origin around 6 million years ago, they provide a very broad time frame for when the host-switch could have possibly occurred, sometime 2–3 million years ago to 10,000 years ago. But they propose that the host-switch likely took place between ten to hundreds of thousands of years ago. Thus, they find their host-switch hypothesis compatible with Malaria's Eve. First, the lateral transfer occurred between chimpanzees and humans, giving rise to *P. falciparum* s.l.; later, a bottleneck and population expansion occurred, giving rise to *P. falciparum* s.s. represented in contemporary human populations.

These findings also offer an explanation as to why Old World monkeys are not susceptible to *P. falciparum*, whereas New World monkeys are. As previously described, Martin and colleagues (2005) found a human-like pattern of Neu5Ac predominance in owl monkeys. A number of New World monkey species other than *Aotus* have been infected with *P. falciparum* under experimental conditions, including *Alouatta palliata* (Baerg and Young 1970; Rossan and Baerg 1975; Taliaferro and Taliaferro 1934), *Ateles fusciceps* (Baerg and Young 1970), *Saimiri boliviensis* (Campbell et al. 1986; Whitely et al. 1987), *Saimiri sciureus* (Campbell et al. 1986; Gysin, Hommel, and da Silva 1980), and *Saguinus geoffroyi* (Porter and Young 1967). The experimental infections represent platyrrhines from four different families: Callitrichidae (*Saguinus*), Cebidae (*Saimiri*), Aotidae (*Aotus*), and Atelidae (*Alouatta* and *Ateles*). This suggests that susceptibility to *P. falciparum* may be quite widespread in New World monkeys. In addition, Duarte and associates (2008) recently found isolated cases of *P. falciparum* in two species of howler monkeys in the wild, apparently anthroponotically acquired from humans.

We understand *how* New World monkeys (or at least *Aotus nancymaae*) are susceptible to *P. falciparum* but not *why*. The selective pressures that led to the parallel evolution of a sialic acid condition that has created an analogous reichenowi resistance/falciparum susceptibility is a puzzle. New World monkeys diverged from Old World primates at least *30 million* years ago (for example, Rosenberger, Hartwig, and Wolff 1991; Takai et al. 2000; see also Cormier 2003 for a review of the platyrrhine fossil record). Even if the human mutation that created susceptibility to *P. falciparum* arose as early as 5 million years ago, platyrrhine monkeys had already been in the New World and separated from other primate species for at least 25 million years. Thus, the mutation that makes New World monkeys susceptible to *P. falciparum* had to have arisen because of selective pressures unrelated to *P. falciparum*.

Since New World monkeys are not naturally parasitized by any form of falciparum-type malaria, it leads to the question: why did the human-like sia mutation arise? If its selection is unrelated to malaria, is it possible that the human mutations related to malaria also arose under unrelated selective pressures? Although little research has been done on the disease course of *P. reichenowi* in the great apes, there is no evidence to suggest that it is similar to *P. falciparum* in its virulence. Thus, would selective pressures been strong enough for humans to evolve genetic defenses specific to *P. reichenowi*? If so, why is it that neither chimpanzees nor gorillas evolved similar defenses to *P. reichenowi*? It is possible that the mutation that provided resistance to *P. reichenowi* in humans arose under selective pressures unrelated to malaria? Asian apes, like Old World monkeys, do not appear to be susceptible to *P. falciparum* infection. The only known documented case of infection is in a splenectomized gibbon (*Hylobates lar*) (Gould and Cadigan 1966). It appears that falciparum-type malaria arose in the primate line in Africa sometime after the divergence of African and Asian apes.

The distribution of vivax-type malaria is even more puzzling. Outside of a few geographically isolated primate species, it does not occur in tropical Africa (with the possible exception of the unresolved status of *P. schwetzi*). It has a similar disparate distribution in South America, occurring in only a few species and failing to spread to close relatives, unlike the malariae-type *P. brasilianum,* which occurs throughout tropical South America in at least 34 species. Although vivax-type malaria is rare in cercopithecine and colobine monkeys in Africa, it is highly diversified in cercopithecine and colobine species in Asia. Questions related to vivax-type malaria are addressed in the next chapter.

BONOBOS AND GORILLAS

The hypothesis that human *P. falciparum* arose from a host-switch from chimpanzee *P. reichenowi* is persuasive in its ability to both explain the origins of human falciparum and explain its virulence. However, subsequent to Rich and associates' (2009) study, evidence has come to light that suggests that bonobos or gorillas could be involved in a lateral transfer to humans. The bonobos (*Pan paniscus*) are closely related to the common chimpanzees (*Pan troglogytes*). For most of the history of anthropological/primatological thought, chimpanzees have been considered to be our closest genetic and behavioral analogues. The bonobo, although just as closely related, has been less studied in the wild.

Krief and associates (2010) have proposed that human falciparum evolved as a host-switch from bonobos. They sampled mitochondrial genomic sequences from 49 chimpanzees and 42 bonobos and found a malaria parasite in the bonobos that clustered with *P. falciparum*. However, they found that the *P. cf. falciparum* mitochondrial lineages derived from bonobos (four in subpopulation) were 10 times more diverse than that of all human *P. falciparum*. They believe this evidence supports a host-switch of *P. falciparum* from bonobos to humans. Another important finding by Krief and associates (2010) was the identification of two additional falciparum-type malarias in chimpanzees, *P. billbrayi* and *P. billcollensi*, which they have assessed to be distinct from the *P. gaboni* identified by Ollomo and colleagues (2009).

Gorillas have also been found to harbor *P. falciparum* (Duval et al. 2010; Liu et al. 2010; Prugnolle et al. 2010). Using primarily fecal material, Prugnole and associates (2010) conducted cytochrome b studies 125 chimpanzees and 84 gorillas. They found that western gorillas harbor not only *P. falciparum* but also two previously unknown lineages of falciparum-type parasites, which they designated *P. GorA* and *P. GorB*. Liu and associates (2010) conducted a similar study using fecal material but with a large sample size of approximately 1,800 chimpanzees, 800 gorillas, and 100 bonobos. They found that human *P. falciparum* formed a monophyletic clade within the gorilla *P. falciparum* radiation and concluded that human falciparum likely originated as a lateral transfer from gorillas. For chimpanzee parasites, Liu and associates (2010) found that none was closely related enough to human *P. falciparum* to be the progenitor (see also Hughes and Verra 2010; Ricklefs and Outlaw 2010; Silva et al. 2010). Regarding the findings of Krief and colleagues (2010) on bonobo falciparum, the authors suggest that the four in their sample had anthroponotic human falciparum infections owing to being housed in an urban sanctuary (see also Hughes and Verra 2010; Silva et al. 2010).

One additional finding of Liu and associates (2010) was a pattern of finding plasmodium infection in western great ape species but not in eastern great ape species. A similar west/east pattern of vivax-type malaria parasites is found among African monkeys. This is further discussed in Chapter 4 and the possibility is raised that climatic/ecological factors may be at play with a west African Pleistocene refuge. In

South America, it is the mirror image with vivax-type malaria occurring in Neotropical monkeys in the east but not the west. However, in the case of New World monkeys, it will be argued that vivax-type malaria (*P. simium*) likely has a disjunct distribution owing to deforestation over the last several hundred years.

In conclusion, the evidence suggests that humans acquired falciparum malaria as a host-switch from a great ape species. The timing of the host-switch remains unclear, but it is likely that questions will soon be resolved with advances in molecular biology. We have learned more about falciparum-type malaria in the last two years than in the last 50. Regardless of the time of the original great ape host-switch, the position taken in this work is that in considering multiple lines of evidence, epidemic falciparum malaria in humans is associated with the rise of agriculture. A further important implication of a great ape host-switch is that one or more great apes may potentially serve as a reservoir for human malaria, seriously compromising efforts at malaria eradication in human populations. In addition, as Duval and associates (2010) have noted, the potential also exists for virulent human falciparum to infect great apes, threatening their existence in the wild.

NOTES

1. *Plasmodium* is actually paraphyletic, because *Hepatocystis* and *Haemoproteus* are included within the clade of plasmodium (see Perkins and Schall 2002).
2. More conservatively, Hagner and associates (2007) analyzed the SSU rRNA, cytochrome b, and the plastid caseinolytic protease C (ClpC) gene and have argued that the data presently available cannot resolve the phylogenetic origin of *P. falciparum*.
3. Livingstone (1971) also suggested that *P. malariae* may have arisen as a lateral transfer from chimpanzees.

Chapter 4

Vivax Type: The Macaque Malaria

Vivax malaria in humans is *typically* a much more benign form than falciparum malaria, but it is much more widespread. It is estimated that between 130 and nearly 400 million clinical infections occur each year and that 40% of the world's population (~2.6 billion people) is at risk from human vivax (Price et al. 2007). While it is a significant disease in the Middle East and some parts of Africa, it is the *major* cause of malaria outside Africa, mainly afflicting peoples of Asia and South America (Carlton et al. 2008; Carter and Mendis 2002). Historically, it was widespread in North America and Europe, before being largely eradicated (Humphreys 2001; Sallares, Bouwman, and Anderung 2004; see also Chapter 1). Although vivax is often considered a benign form of malaria, malignant vivax has been documented in Papua New Guinea (Anstey et al. 2009). However, even the relatively benign expression of vivax in comparison with human falciparum can create serious health problems, particularly in children, including malnutrition, anemia, growth retardation, and stunted physical development (Price et al. 2007). Complications from human vivax malaria include splenic rupture, thrombocytopenia, renal failure, and shock (Anstey et al. 2009). Human vivax also imposes a tremendous economic burden: estimates of the costs from treatment and loss of work range from U.S. $1.4 to 4.0 billion per year (Price et al. 2007).

The question of the age, origin, and spread of vivax-type malarias has involved more debates and puzzles than have questions regarding falciparum-type malarias. At least for the falciparum-type malarias, there is agreement that they originated in Africa and that human and great ape falciparum-type malarias are either sister taxa or the result of host-switches. Understanding the history of the vivax-type malarias has presented a much more complicated picture with enigmas that remain difficult to resolve. In many cases, the scenarios suggested by molecular genetics have been at odds with the anthropological evidence concerning human migrations and wild-primate distributions. Vivax-type malaria is designated here as the "macaque malaria," because Southeast Asia appears to be a center of endemicity with at least eight different species affecting wild primates, and because this type of malaria is particularly prominent in macaques.

Regarding the origin of vivax-type malarias, two main camps exist, "Out of Asia" and "Out of Africa." In brief, the out-of-Asia hypothesis argues that the vivax-type malarias originated in Southeast Asian wild primates and spread relatively recently to humans through a host transfer, perhaps within the last 50,000 years. The out-of-Africa hypothesis holds that vivax malaria originated in Africa and is likely older than falciparum-type malaria. In addition, it has even been proposed that vivax-type malarias originated in New World monkeys and spread from Amazonia to Old World human and wild-primate populations within the last 500 years. While the latter scenario seems unlikely, evidence that could support such a view is addressed in greater depth in the next chapter. In addition, I make the case that the primate fossil record suggests one additional possibility: that vivax-type malaria could have entered Asia among any number of hominoid species and could have been maintained through host-switches among ancient hominins and other hominoids before becoming established in its contemporary forms in humans and Asian wild primates.

General agreement exists that human *P. vivax* and New World monkey *P. simium* are sister taxa and that a host-switch occurred, in some direction, between humans and New World monkeys. Studies are also in agreement that there is a very close relationship between vivax malarias of humans and Asian/Southeast Asian wild primates, suggesting that a host-switch occurred between humans and wild primates here as well, in some direction. What is not clear is if the original host of vivax-type malaria is human or monkey (or possibly Asian ape) or if vivax type first arose in Africa or Asia. Both the out-of-Asia and out-of-Africa views have difficulty in explaining the disjunct distribution of *P. simium,* found in New World monkeys of the Atlantic coastal forest, and, as well, the disjunct distribution of *P. gonderi*, found in monkeys on the west coast of Africa.

Apart from human vivax malaria, a few cases of *P. vivax* identified in chimpanzees, and the relatively isolated distributions of vivax-type malaria in east coast South America and west coast Africa, all other primate forms of vivax-type malaria are found in Asia and Southeast Asia. It remains a puzzle as to how *P. simium* and *P. gonderi* came to be established in geographic areas so distant from Asia and Southeast Asia, and it is equally puzzling that these two forms appear to have failed to spread more widely to other wild primates in South America and Africa.

This chapter explores the origins of vivax-type malaria through both an examination of the contemporary distribution of primates infected with vivax-type malaria and the potential species affected from the fossil record.

Molecular Genetics and the Origin of Vivax Malarias

Many of the same difficulties found in constructing molecular phylogenies with falciparum-type malaria parasites are also true of the vivax group but are compounded, for less is known about the basic biology and genomic diversity of *P. vivax* than that of *P. falciparum* (see Carlton et al. 2008; Feng et al. 2003), and even less about the vivax-type parasites of wild primates. Depending on the particular gene loci under analysis, somewhat different trees have emerged. Although the relationships among the vivax malarias have not yet been completely resolved, it is has become very clear that all share a close genetic relationship and are much closer to one another than they are to any of the other human or nonhuman primate malaria types (falciparum, malariae, and ovale) (for example, Ayala et al. 1998; Ayala, Escalante, and Rich 1999; Cornejo and Escalante 2006; Escalante, Barrio, and Ayala 1995; Escalante et al. 1998; McCutchan et al. 1996; Mu et al. 2005; Perkins and Schall 2002; Qari et al. 1996). Apart from human *P. vivax,* the members of the vivax-type clade in wild primates are listed below (see Appendix 1 for a complete reference list).

The best known members of vivax-type malaria group are human *P. vivax,* the closely related New World monkey *P. simium*, and the plasmodia that infects Asian monkeys, particularly the macaques. The gibbon and orangutan malarias also likely belong to the vivax group, but most have not been well studied through genetic analysis. The gibbon malaria, *P. hylobati,* is often listed with the vivax group in the literature, and one research group (Steiper et al. 2005) is now classifying the orangutan parasites of *P. pitheci* and *P. silvaticum* as members of the vivax type based on molecular analysis.[1] The remaining Asian ape plasmodia that infects gibbons (*P. eylesi, P. jefferyi,* and *P. youngi*) are suspected to belong to the vivax group based on morphology and geographic distribution in Southeast Asia. Similarly, among the African monkey forms of vivax-type malaria, only *P. gonderi* has been genetically determined to belong to the vivax group. The other two forms found in African monkeys, *P. georgesi* and *P. petersi,* are suspected to belong to the vivax group due owing their species and geographic distributions (Gysin 1998; Poirriez et al. 1995).

The status of African ape *P. schwetzi* remains unclear but is believed to belong either to the vivax type or the ovale type. Resolving the status

Table 4.1 Vivax-Type Malaria in Wild Primates

Primate Group	Plasmodia Species	Primate Species
Asian Monkeys	*P. coatneyi*	*Macaca fascularis*
	P. cynomolgi	*Macaca arctoides, M. cyclopis, M. fascicularis, M. mulatta, M. nemistrina, M. radiata, M. sinica, Trachypithecus cristatus, Semnopithecus entellus*
	P. fieldi	*Macaca fascicularis, M. nemistrina*
	P. fragile	*Macaca radiata, M. sinica*
	P. inui	*Macaca assamensis, M. cyclopis, M. fascicularis, M. mulatta, M. nemistrina, M. nigra, M. radiata, M. sinica, Presbytis melalophos, P.* sp., *Trachypithecus cristatus, T. obscurus*
	P. knowlesi	*Macaca cyclopis, M. fascicularis, M. nemestrina, M. nigra, Presbytis melalophos*
	P. simiovale	*Macaca sinica*
Asian Apes	*P. eylesi**	*Hylobates lar*
	P. hylobati	*Hylobates moloch, Hylobates* spp.
	*P. jefferyi**	*Hylobates lar*
	*P. pitheci**	*Pongo pygmaeus* s.l.
	*P. silvaticum**	*Pongo pygmaeus* s.l.
	*P. youngi**	*Hylobates lar*
African Monkeys	*P. georgesi**	*Cercocebus agilis, Lophocebus albigena*
	P. gonderi	*Cercocebus atys, Cercocebus agilis, Cercocebus* sp. undet., *Lophocebus aterrimus, Mandrillus leucophaeus*
	*P. petersi**	*Lophocebus albigena*
African Apes	*P. schwetzi***	*Gorilla gorilla* s.l., *Pan troglodytes*
	P. vivax	*Pan troglodytes*
	P. cf. vivax	
New World Monkeys	*P. simium*	*Alouatta caraya, A. guariba, Brachyteles arachnoides*

*Presumed, but not verified, to belong to vivax type based on morphology, geographic distribution, and species distribution.
***Plasmodium schwetzi* on morphological and life-history characteristics is believed to belong to either the vivax type or an ovale type.
See Appendix 1 for reference list.

of the phylogenetic relationship of *P. schwetzi* to other malarias will be extremely important in understanding the relationships among the primate malarias. It has been identified in gorillas (Bray 1963; Coatney 1971; Coatney et al. 1971; Contacos and Coatney 1963; Warren and Wharton 1963) and chimpanzees (Bray 1958b, 1963; Coatney 1971; Coatney et al. 1971; Contacos and Coatney 1963; Warren and Wharton 1963). If *P. schwetzi* ultimately proves to be a member of the ovale group, it will represent the only other member of ovale-type malaria, beyond *P. ovale*. If it ultimately proves to be a member of the vivax group, it will be extremely valuable in understanding the puzzling distribution of vivax-type malarias in Africa. Note also that both *P. vivax* (Kaiser et al. 2010; Krief et al. 2010) and *P. ovale* (Duval et al. 2009, 2010; Kaiser 2010) have recently been identified in chimpanzees, but it is unclear if chimpanzees are natural hosts for these two parasites, or if they acquired the infections through anthroponotic transfer.

Among the known members of vivax-type malaria clade that have been analyzed and sequenced, their genetic relationships are still far from clear—meaning that although we know that all these vivax forms are related to one another, to date it has not been established with certainty, based on current evidence from molecular phylogeny, which plasmodial species are related through sharing a common ancestor and which are genetically similar owing to lateral host transfer. The one notable exception is the relationship between human *P. vivax* and New World monkey *P. simium;* they are so closely related that some malaria researchers suggest they might best be considered indistinguishable conspecifics (for example, Escalante, Barrio, and Ayala 1995; Ewald 1983; Fandeur et al. 2000; Lal et al. 1988; LeClerc et al. 2004; Rich and Ayala 2000). In this view, *P. vivax* and *P. simium* actually represent a single species of plasmodium, but it is designated as "vivax" when occurring in humans and as "simium" when occurring in New World monkeys.

Such multihost distribution is in keeping with the pattern seen in Asian monkeys. However, Asian monkeys are much more closely related to one another than are humans and New World monkeys; moreover, Asian monkeys share a much greater time depth of ecological association than do humans and New World monkeys, on the scale of 10 million years (Pan, Groves, and Oxnard 2004; see also following discussion), contrasting with approximately 10–15 thousand years of human and New World monkey association. Because humans and New World primates are both genetically distant and have been geographically separated for at least 30 million years, plasmodial sharing can be explained only by a host-switch, in some direction, either when humans first colonized South America, approximately 10,000–15,000 years ago, or with the European colonization, approximately 500 years ago (see Chapter 5).

Thus, the *P. vivax* and *P. simium* host-switch took place with more distantly related primate species and likely within a much more recent time frame than potential Asian primate host-switches.

With the exception of *P. vivax,* the only genetically verified member of vivax-type malaria to date that occurs naturally in tropical African primates (human or wild) is *P. gonderi*, present in a few species of mangabeys and drills in a geographically circumscribed area extending from the western/central coast. Even if the African mangabey *P. georgesi* and *P. petersi*, as suspected, are proven to be members of vivax-type group, the distribution of African tropical vivax-type malaria would be little changed. The mangabeys affected with *P. georgesi* and *P. petersi* occur in the same general geographic region as the mangabeys and drills affected with *P. gonderi*. In the literature, *P. gonderi* is usually treated as an outgroup to the Asian forms, due to its geographic distance. However, in a study by Roy and Irimia (2008), they found a deletion in the mitochondrial large subunit rRNA specific to *P. gonderi*, *P. cynomolgi* and *P. simiovale.*[2] It was not found in the other primate vivax parasites they examined: *P. fragile, P. knowlesi*, *P. vivax*, and *P. simium*. They suggest that the genetic difference may warrant considering gonderi, cynomolgi, and simiovale as members of a subclade within the vivax group. Such a genetic relationship is difficult to explain in terms of geography, ecology, and history. If the possibility exists that African *P. gonderi* is not a distant outgroup to Asian vivax type but rather shares a common ancestor with some, but not all, Asian vivax forms, we truly have a difficult puzzle to resolve in putting together evolutionary history, biogeography, and primate migration and in attempting to determine the ultimate geographic and species origins of vivax-type malaria.

OUT OF ASIA OR OUT OF AFRICA?

The debate as to whether vivax-type malarias originated in Africa or Asia (or possibly even the wildcard view that vivax type originated in the New World) has implications regarding potential co-evolutionary adaptations, genetic diversity, and geographic distribution of concern to medical anthropologists, primatologists, and geneticists. Broadly, the out-of-Asia view is based on both the diversity of the vivax-type plasmodia and the diversity of wild Asian primates affected. Eight known plasmodia species in Asian wild primates are classified as belonging to the vivax group: seven in monkeys and one in the gibbon. It is likely that five more Asian species will be classified as members of the vivax group: two in orangutans and three in gibbons. At least 15 wild-primate species are naturally infected with vivax-type malarias; in many cases, the same plasmodium species infects multiple wild-primate species, and the same primate species is infected by multiple plasmodia species. Thus Asia, and particularly Southeast Asia, can be considered a center of endemicity for vivax-type malaria. The speciation of Asian monkeys likely plays a role in the speciation of Asian plasmodia, but a number of host-switches may very well have occurred, which would help to account for the lack of host-specificity for many of the monkeys. (This is addressed again later on.)

Although Southeast Asia is a center of endemicity for vivax-type malaria, this does not necessarily mean that it is the point of origin for primate vivax-type malaria. One study that set out to specifically test the hypothesis of origin compared the mitochondrial genomes of 176 human *P. vivax* isolates from various regions of the world (Mu et al. 2005). This study found that the malaria parasite populations in Melanesia, Southeast Asia, and India appeared to be more diverse than those in Africa, South America, and Central America. If that is the case, it suggests that *P. vivax* has been evolving longer in human populations in Asia than in Africa or the New World. The researchers interpreted their results to suggest that *P. vivax* likely became a human parasite after a host-switch from an Asian macaque. In addition, they estimated the most recent common ancestor of *P. vivax* to have arisen 53,000–265,000 years ago, using the divergence of macaque and baboon ancestors as a root.

In a similar study, Cornejo and Escalante (2006) compared complete mitochondrial genomes of human *P. vivax* in several regions. They found the greatest regional haplotype diversity in Asia, followed (in rank order) by Melanesia, India-Middle East, Africa, and the Americas. The findings were interpreted to mean that the greater diversity of the Asian haplotypes suggested that it was older. They also interpreted their results as supporting an origination of *P. vivax* in Asia, probably as a host-switch from a macaque. In addition, their molecular clock estimates suggested that the time for the most recent common ancestor of *P. vivax* could be as old as 460,000 years, considerably earlier than the estimates of Mu and associates (2005).

As a brief aside, the molecular clock hypothesis is an inferential statistical technique that is used to estimate the time when two species diverged from each other, but many problems still exist with the precision of its timing (see reviews of Benton and Donoghue 2006; Drummond et al. 2006; Hedges and Kumar 2003). The molecular data that is used is typically nucleotide sequences from DNA or RNA or amino acid sequences from proteins. The underlying assumption is that consistency exists in rates of mutation over time. Very simply put, it estimates how long ago two species diverged from each other looking at how many changes can be seen in nucleotide and amino acid sequences. Two of the most important issues relating to the accuracy of molecular clocks for plasmodia are accurate calibration and accurate determination of rates of change. Errors in either will lead to the clock not keeping good time. Calibration involves setting a correct time to start the clock in motion. If the rate is accurate but the date of calibration is wrong, then the information about the relationship between species is correct but the timing of speciation is wrong. Molecular clocks are often calibrated by using the fossil record. One advantage in using the molecular clock for the primate malarias is that the primate fossil record has been particularly well studied in comparison to other species.

One of the main arguments in favor of an African origin of *P. vivax* involves the presence of RBC Duffy negativity among indigenous peoples of tropical Africa. Both its prevalence in tropical Africa and its patterns of distribution have suggested that vivax-type malaria may have originated in Africa and have had a great time depth on that continent. As described in Chapter 2, RBC Duffy negativity is a homozygous state of the RBC antigens that is at near fixation in tropical Africa and confers complete immunity to *P. vivax* malaria. The frequency of RBC Duffy negativity is over 97% in the indigenous populations of sub-Saharan and West Africa, 70% in other parts of Africa, and 50–60% in some parts of the Arabian peninsula (Carter 2003; Carter and Mendis 2002). Given the pervasiveness of RBC Duffy negativity in tropical Africa, its decreasing frequency in other parts of Africa and the Mediterranean, and rarity elsewhere, researchers have suggested that it is an adaptation against vivax-type malaria. As RBC Duffy negativity decreases, the frequency of vivax-type malaria in humans increases.

Vivax-type malaria has also been argued to have a greater time depth than falciparum-type malaria in humans, because the potential genetic adaptations for *P. falciparum* confer a degree of resistance only while RBC Duffy negativity provides complete immunity to *P. vivax*. Thus some have interpreted it to mean that co-evolutionary processes between humans and *P. vivax* have been at play longer than those between humans and *P. falciparum* (for example, Carter and Mendis 2002; Miller et al. 1976). The issue of virulence is also relevant, not only because of the view that the relative benignness of *P. vivax* may indicate longer co-evolution but also because if it were benign in the past, it would have taken longer for RBC immunity to develop against a less virulent pathogen. In this case, the suggestion is that *P. vivax* originated in Africa, but genetic adaptations developed to eliminate it from human populations. But as discussed at length in Chapter 2, virulence is not a reliable indicator of age.

Although it is clear that RBC Duffy negativity confers protection against human vivax malaria, the evidence that it was selected for as an adaptation against vivax malaria is only circumstantial. RBC Duffy negativity could have evolved for reasons unrelated to malaria resistance in African populations. If RBC Duffy negativity arose for reasons unrelated to *P. vivax*, this could have two implications. First, *P. vivax* may still have originated in Africa but was only incidentally eliminated in populations that developed RBC Duffy negativity mutation. Second, *P. vivax* may have originated outside Africa but could not become established in populations with RBC Duffy negativity.

Another potential factor that is not often considered is that biogeographic factors may play an important role in the absence of *P. vivax*-type malaria in human populations in tropical Africa. The distribution pattern of vivax-type malaria in wild primates would seem to support that view. As described, the only definite member of the vivax group in Africa, apart from *P. vivax*, is *P. gonderi*. Unfortunately, we do not have exact geographic information on the original monkeys from whom gonderi was isolated, except that they were in Cameroon, Liberia, and the lower Congo (Coatney et al. 1971; Garnham 1963). However, in combination with the contemporary distributions of the primate species involved, it seems that *P. gonderi* is restricted to the closely related the drill and mangabey species near the central west coast of Africa. The same applies to the suspected vivax-type malarias in Africa, *P. georgesi* and *P. petersi*.

Mangabeys have now been separated into two genera, the arboreal *Lophocebus* and the terrestrial *Cercocebus*; further, *Cercocebus* species have been found to be more closely related to the terrestrial drill (*Mandrillus leucophaeus*) than they are to the *Lophocebus* "mangabeys" (McGraw and Fleagle 2006). One the one hand, since *Cercocebus-Lophocebus-Mandrillus* form a clade, their susceptibility to vivax-type parasites could arguably have evolved independently in the basal ancestor that gave rise to the clade. On the other hand, there are several reasons to question whether or not that is the case. One is that if a form of vivax were specific to that clade, the expectation would be that other species in that clade, but outside western coastal Africa would be affected by vivax-type malaria. *Cercocebus* species extend eastward into the Congo basin and are present as far east as the Tana River, near the east coast of Africa (International Union for Conservation of Nature and Natural Resources 2009). It is possible that vivax malaria does exist in other mangabeys but has not yet been documented. A bigger issue is that although the mangabeys and the drill belong to the same subfamily as the macaques, Cercopithecinae, the vivax malaria parasites also infect members of the Colobinae subfamily in Asia. However, we have not yet found evidence of vivax-type malaria parasites in any other African cercopithecines or colobines.

If biogeographical factors are at play, they do not seem to be related to distribution in the canopy. *Cercocebus* species and *Mandrillus* are both largely terrestrial, whereas members of the *Lophocebus* species are largely arboreal. Thus, their susceptibility does not seem to be related to niche specialization vertically in the canopy (which could implicate habitats of mosquito vectors) but occurs more generally in the tropical African west coast. The unclassified species that have been identified in mangabeys (*P. petersi* and *P. georgesi*) are likewise restricted to the same general area. In addition, as noted before, the distribution of *P. simium* in South America is quite similar and is restricted to only a few species in the Atlantic coastal forest. It provides almost a mirror image of the African distribution. Looking at Pangaea, the distribution of *P. simium* and *P. gonderi* appears continuous, but that is not a possible explanation, because Pangaea separated around 250 million years ago, and playtrrhine monkeys arrived in South America around 30 million years ago.[3]

In South America, *P. simium* has been documented in only a few Atlantic coast forest wild primates. However, the range of one of the howler species that serves as a natural host, *Alouatta caraya,* extends considerably westward across the continent, and other howler species are found throughout Amazonia. Given that, it seems the limiting factor in the spread of *P. simium* in South America may be ecological in nature rather than based on genetic susceptibility. In contrast, we do not find this kind of ecological circumscription in New World monkey *P. brasilianum* (malariae type), because it is widely distributed in New World monkeys. An additional potential confounding factor that should be considered is that the geographical areas in the African and South American continents where *P. gonderi* and *P. simium* occur in wild primates are both highly disturbed areas, so human-mediated factors may be involved. One possible explanation is that these coastal anthropogenic environments may be particularly favorable to conditions for vector-host interaction and the proliferation of malaria. Another possible explanation is that the critical anthropogenic factor is that coastal areas, as ports of call, may have created cultural conditions that fostered lateral transfer of plasmodia. Meaning, the presence of vivax-type malaria in wild primates in these restricted zones may be due to recent events involving colonization and trade, whereby vivax-bearing humans (or their pet primates) introduced vivax-type malaria to new species, rather late in the game from an evolutionary perspective.

One of the most elaborated models of the origin of *P. vivax* comes from Richard Carter (2003).

His work incorporates Pleistocene refuge theory and the findings by Li and associates (2001) that vivax-type malaria was introduced twice into the New World. First, the study by Li and associates (2001) found derived differences between Asian and New World forms of human *P. vivax*. Surprisingly, they found that Neotropical *P. simium* clustered with the human Asian *P. vivax* variant rather than with the human New World *P. vivax* variant. If the closest relative of *P. simium* is with an Old World human variant of *P. vivax* rather than a New World variant, it would mean that Neotropical monkey *P. simium* and New World human *P. vivax* variant do not share the same immediate common ancestor. Although Li and associates (2001) do not speculate on the origins of the two forms in the New World, they do suggest that the widely held view that *P. vivax* entered the New World only once with the European colonization is not the only possible scenario.

One aspect of Carter's (2003) hypothesis suggests a Pre-Columbian entry of a variant of *P. vivax* into South America from Southeast Asia by boat across the Pacific. In his view, the form of vivax malaria introduced by hypothetical Southeast Asian seafarers failed to become established in indigenous human South Americans but was transferred to monkeys. Later, Europeans colonists harboring another variant of *P. vivax* introduced it to New World indigenous peoples, where it was perpetuated among their descendants. Carter is neither the first nor the only researcher to have raised the possibility that malaria existed in humans and/or monkeys before European colonization. However, at this point it should be stressed that a scenario of a Southeast Asian people traveling to South America is currently at odds with evidence from cultural anthropology, biological anthropology, linguistics, and archaeology. As is discussed in the next chapter, remarkably, it appears that Pre-Columbian contact did occur between Remote Oceania and South America, a truly vast distance to traverse with low-technology watercraft. But there is no evidence to suggest that Pre-Columbian malaria existed in Remote Oceania. Boat travel from by the nearest trans-Pacific route of plasmodia-bearing humans would be from Southeast Asian populations, effectively doubling the distance from Remote Oceania and posing logistical issues that are difficult to justify. Even if, conceivably, Southeast Asian humans bearing malaria traveled to Remote Oceania in Pre-Columbian times, theoretically introducing it to Remote Oceania peoples, there were no anopheline mosquitoes present in that region at that time to establish a host-vector cycle. But more to the point here, Carter proposed an out-of-Africa/Mediterranean scenario, circa 10,000 years ago, that involved an eastward and westward movement of two strains of *P. vivax* that had been isolated in Pleistocene refuge areas.

Pleistocene refuge theory offers a potential explanation for species diversification in the tropics owing to climate and vegetational changes. The earth's history is characterized by five major "ice ages." Ice ages are characterized by periods of relatively colder and warmer periods, glacials, and interglacials. Although the term "ice age" is sometimes loosely used as synonymous with glacials, we are actually in the last ice age, the

Quaternary, in an interglacial period. The Quaternary Ice Age began approximately 2.5 million years ago at the beginning of the Pleistocene, and the current interglacial began with the Holocene around 12,000 years ago (Augustin et al. 2004; Hewitt 2000). According to Pleistocene refuge theory, before the Quaternary ice age, tropical forest organisms lived in continuous forest; cooling temperatures during the glacials led isolation of some species into warmer forest refuges, where they speciated by means of genetic drift (Mayr and O'Hara 1986).

Carter (2003) suggests that the ancestor of *P. vivax* likely diverged from an ancestral vivax-type malaria roughly around the start of the Quaternary Ice Age, 2–3 million years ago. Because primate species were geographically widespread in Africa and Eurasia, he posits that the geographic place where ancestral *P. vivax* diverged could have been anywhere in the Old World. However, as temperatures cooled and forests fragmented (according to refuge theory), anopheline mosquitoes were circumscribed to refugia in sub-Saharan Africa, the Mediterranean, and Southeast Asia. While human ancestors and related species could live outside these areas, he argues that the malaria-causing plasmodia went extinct except in the refuges. He also argues that if ancestral *P. vivax* was present in the Southeast Asia refuge area, it likely went extinct in here because of the absence of fossil evidence in hominins 100,000 to 60,000 years ago in Asia and Southeast Asia. This last point does not appear to be the case, which will be discussed further below.

Carter (2003) then offers an explanation for the spread of the two variant forms of *P. vivax* malaria proposed by Li and associates (2001). When temperatures began to warm in the Holocene, he suggests that the *P. vivax* variant in sub-Saharan Africa began moving east, entered Asia by land and the western Pacific by sea with human boat travel. From there, humans infected with vivax traveled from Oceania to South America where vivax entered the New World in Pre-Columbian times. New World monkeys were then infected by lateral transfer from humans. Regarding the Mediterranean *P. vivax* variant, he proposed that it was originally an infection of Neanderthals that spread west and north by means of a lateral transfer to modern humans as they migrated out of Africa. The presumed Mediterranean variant persisted in human populations in Europe and was eventually carried into the New World with colonization 500 years ago, entering into the indigenous populations.[4]

One strength of Carter's model is that it could provide a biogeographical explanation for the limited distribution of *P. gonderi* in African monkeys. It should be mentioned that Carter is defining these potential "refugia" very broadly. The exact locations of refuges are still controversial and very much debated. Nonetheless, a recent analysis of gorilla genetic structure found a demographic expansion of eastern lowland and

Gabon/Congo mitochondrial haplogroups that were consistent with a post-Pleistocene expansion from a formerly isolated refugia (Anthony et al. 2007). What is significant about the gorilla analysis is that the area identified as a Pleistocene refuge roughly corresponds with the areas of distribution of the mangabey and drill species where *P. gonderi*, *P. georgesi*, and *P. petersi* have been identified. In addition, as discussed in Chapter 3, it also seems that falciparum-type malaria follows a similar eastern distribution among the African great apes.

Although the origin of human vivax is yet unresolved, at this point we have two major scenarios for the origin and spread of vivax-type malaria that offer explanations for the disjunct distribution of the existence of the genetically determined *P. gonderi* vivax-type malaria in Africa. Several subscenarios correspond, but at this point we will just look at the premises of the two major explanations. One is that the progenitor of vivax-type malaria is quite ancient and was widespread in Africa. It would have been old enough to have accompanied Old World monkeys when they migrated to Eurasia 10 million years ago. Beginning with the Quaternary Ice Age, vivax-type malaria began to be circumscribed into warmer areas. By the time of the Pleistocene, vivax-type malaria survived only in warmer refuge areas. One of hypothesized refuge areas, based on *Gorilla* systematics and biogeography, is posited to be in the region of coastal West Africa, in the same general vicinity of vivax-type *P. gonderi* as well as unclassified mangabey.

P. georgesi and *P. petersi*

The second major scenario offers very different premises. It suggests that the ancestral vivax-type malaria arose in situ in Southeast Asia, or perhaps somewhere else in Asia. It diversified among Asian monkeys and apes through a combination of lateral transfers and codivergence with speciations among Asian primates. When human ancestors or relatives entered Asia, a form of vivax-type malaria was introduced to hominid/hominin populations from a lateral transfer from wild primates. The presence of the vivax-type *P. gonderi* in contemporary African monkeys (and likely *P. petersi* and *P. georgesi*) here would be due to a relatively recent lateral transfer of vivax-type malaria from humans at coastal West African ports of call. Here, humans may have introduced vivax-type malaria directly to coastal African monkeys, or it may have been introduced indirectly through lateral transfer from Asian/Southeast Asian malaria-bearing primate pets of humans. Although such a scenario might initially seem more far-fetched, its merit is bolstered, because it mirrors the distribution of vivax-type malaria in South American monkeys. Like the African case, in South America, the distribution of vivax-type malaria is limited to relatively few species, existing in a coastal port of call region, which failed to disperse more widely among members of the same genera.

The primate fossil record in Africa and Eurasia is a key source of information needed to understand the genetic analysis. Evidence from the fossil record raises an additional possibility that vivax-type malaria could have been maintained in Asia and Southeast Asia in *hominoids* for perhaps as long as 10 million years. If this is the case, it in no way negates a recent host-switch in some direction from cercopithecoid monkeys and humans but suggests that Asian hominoids from Miocene apes to ancient hominins may play a more important role in understanding vivax malaria than previously suspected.

A brief review of the primate fossil record follows, including its implications for understanding the history of vivax-type malaria.

The Primate Fossil Record

Over the last 20 years, molecular genetics has become increasingly important in elucidating primate evolutionary history. Nonetheless, examination of the fossil record still has a critical place in understanding evolution. One advantage of examining the fossil record is that can provide geographic locations for primate ancestors and can elucidate migration patterns, which are particularly relevant for understanding the relationship among the vivax-type malarias.

At present, the molecular analyses supporting an Asian origin of vivax malaria (and presumed host-switch from monkeys to humans) provide a very broad time frame within which this would have occurred. Depending on the study, it has varied from 10,000 years ago to over 400,000 years ago. Carter (2003) puts the time frame for the *expansion* of human vivax malaria within the last 10,000 years but places the time of divergence of ancestral *P. vivax* as early as 2–3 million years ago. An examination of the Old World primate fossil record has potential to clarify when and where primates bearing malaria-causing plasmodia came into contact and how host-switches may have occurred.

Fossil Apes

Among the higher primates in the Old World, one of the best known early fossils is *Aegyptopithecus zeuxis* from the early Oligocene (29–30 million years ago) (for example, Ankel-Sions, Fleagle, and Chatrath 1998; Fleagle and Simons 1982; Simons et al. 2007). It is from an area in Egypt known as the Fayum depression, and although it is considered a higher primate, it actually predates the separation of monkeys and apes. By the time of the early Miocene (23–16 mya), apes (hominoids) and Old World monkeys (cercopithecoids) appear as distinctive groups in Africa. Apes became abundant in the Miocene and diversified with as many as 100 different proposed ape species identified during this epoch (Fleagle 1999).

Today, the best represented ape species are the so-called lesser apes, the Asian gibbons and siamangs (*Hylobates* spp.) with 9–11 members, depending on classification. For the great apes, at the species level, there are only two species of chimpanzees/bonobos (*Pan*), two species of gorillas (*Gorilla*), and two species of orangutans (*Pongo*). Including humans, there are now less than 20 contemporary hominoid species, reflecting only 20% of hominoid representation in the Miocene. With the exception of humans, all the contemporary apes are endangered, with many critically endangered (International Union for Conservation of Nature and Natural Resources 2009). The precarious status of apes today is largely due to habitat destruction; however, it is also true that, with the exception of humans, apes declined after the Miocene, perhaps because of competition with cercopithecoid monkeys.

In the early and middle Miocene (23–11.5 mya), a variety of apes from the Proconsulidae family have been recovered from deposits in from Kenya and Uganda, representing a number of genera, including *Proconsul*, *Rangwapithecus*, *Limnopithecus*, *Dendropithecus*, *Simiolus*, and *Micropithecus*; and from the family Afropithecidae, which includes *Kenyapithecus*, *Equatorius*, and others (Andrews and Kelley 2007; Fleagle 1999; Ward et al. 1999). Although monkeys, apes, and humans originated in Africa and dispersed into Asia by means of a land bridge that connected East Africa and Southwest Asia, it may not be a simple adaptive radiation, because recent evidence suggests that some hominoids may have reentered Africa by the same land bridge.

The Miocene apes were the first anthropoid primates (monkeys and apes) to arrive in Eurasia. Two genera are known from the early Miocene in Eurasia, *Griphopithecus* and *Kenyapithecus*. One of the earliest documented fossil hominoids in Eurasia is c.f. *Griphopithecus*, which comes from the Engelswies site in Germany and dates to 17.0–16.5 mya (Heizmann and Begun 2001); the fossil record of *Kenyapithecus* shortly follows. Both belong to the family Afropithecidae (following the systematics of Moyà-Solà et al. 2009). Two species have been identified from the Palaşar site in Turkey, *Kenyapithecus kizili* and *Griphopithecus alpani*, which date to around 16.5–16 mya (Andrews and Kelley 2007; Kelley 2007; Kelley, Andrews, and Alpagut 2008). A related fossil from the same time frame has been identified from Slovakia and is designated *Griphopithecus darwini* (Andrews and Kelley 2007; Holec and Emry 2003).

A number of apes are known from the Middle Miocene in Europe and Asia (17–10 mya). Several early members of the Hominidae family have been identified in Europe (also following the systematics of Moyà-Solà et al. 2009). Examples from the genus *Dryopithecus* are *Dryopithecus carinthiacus* from Austria, dating to about 12.5 mya (Eronen and Rook 2004; Kordos 2000); *Dryopithecus laietanus* from Spain, dating to about 11.6–9 mya (Moyà-Solà and Köhler 1993); and *Dryopithecus brancoi* from the Rudabánya sites in Hungary, dating to about 10 mya (Armour-Chelu, Andrews, and Bernor 2005; Kordos and Begun 2001). In addition, a new Middle Miocene ape (11.9 mya), *Anoiapithecus brevirostris*, has recently been discovered in Spain and has also been attributed to the family Hominidae (Moyà-Solà et al. 2009).

In addition, several Middle Miocene apes have been documented in Southeast Asia from the Chiang Muan site in northern Thailand. Several molars have been recovered dated to around 11 mya that have been described as having affinities with *Sivapithecus* or *Lufengpithecus* (Kunimatsu et al. 2003; Kunimatsu et al. 2005a; Kunimatsu et al. 2005b). A mandible from the same site has been designated as *Khoratpithecus chiangmuanensis* and is dated to 10–13.5 mya; it appears to be more closely related to *Lufengpithecus* (Chaimanee et al. 2003). A later species, *Khoratpithecus piriyai* from northeastern Thailand, has been argued to be the closest known relative of the orangutan and is dated to 6–9 million years ago (Chaimanee et al. 2004; Chaimanee et al. 2006). But the slightly earlier *Sivapithecus* also has affinities with modern orangutans (Madar et al. 2002; Pilbeam et al. 1990).

Thus, from the fossil record, we have a time frame of apes in Europe 10–12.5 million years ago and apes in Southeast Asia 6–12.5 million years ago. Heizmann and Begun (2001) have hypothesized that the Eurasian hominoids reentered Africa during the late Miocene and may have given rise to the extant hominoid species. They describe evidence extensive exchange of faunal species between Africa and Southwest Asia during the Miocene. *Kenyapithecus* in Turkey dates 1–2 million years earlier than Kenyapithecus at the Fort Ternan site in Africa (Andrews and Kelley 2007). In addition, although apes are abundant in the fossil record in Eurasia during the late Miocene, few are known from Africa. This evidence raises the possibility that contemporary humans and apes may have originally derived from hominoids that evolved for several million years in Eurasia and migrated into Africa. However, as Andrews and Kelley (2007) have pointed out, the scarcity of fossil evidence for late Miocene apes in Africa does not necessarily mean that hominoids were scarce. Although this question cannot yet be answered definitively, the alternative hypothesis that present-day hominids originated from Eurasian hominoids cannot be ruled out.

Fossil Monkeys

Fossils of Old World monkeys are less abundant that those of apes during the early Miocene. One of the earliest Old World monkey fossils that have been identified is *Victoriapithecus* in Kenya that dates to about 19 million years ago, predating the divergence of the two Old World monkey subfamilies: cercopithecine and colobine monkeys (Benefit and McCrossin 1997, 2002). The fossil record suggests that cercopithecine and colobine monkeys diverged in Africa and radiated to Eurasia about 10 million years ago, in the late Miocene or perhaps the end of the middle Miocene (Pan, Groves, and Oxnard 2004; see also Raaum et al. 2005). The living members of the Old World monkey subfamilies, cercopithecines and colobines, are anatomically distinct owing to basic dietary differences with the cercopithecines, specializing on fruits, and the colobines, specializing on leaves and seeds (Fleagle 1999).

In the late Miocene (10–5 mya), primates from the pliopithecid family have also been documented in Eurasia. The pliopithicids are sometimes referred to as the "non-apes" and appear to be an outgroup to both Miocene apes and cercopithecoid monkeys that apparently went extinct during the Pliocene. One of the best known forms is *Anapithecus hernyaki* from the Rudabánya sites in Hungary dated to about 10 million years ago, where they co-occur with the hominoid *Dryopithecus* (Armour-Chelu, Andrews, and Bernor 2005; Kordos and Begun 2001; Nargolwalla et al. 2005). In Spain, *Egarapithecus narcisoi* dates to about 9 mya (Moyà-Solà, Köhler, and Alba 2001).

Old World monkeys were initially abundant in middle Miocene record of Europe but eventually disappeared, with a few groups spreading eastward into Asia between the end of the Miocene and the early Pliocene (Pan, Groves, and Oxnard 2004). Several colobine species of the genera *Mesopithecus* and *Dolichopithecus* have been identified in Europe and Asia. *Mesopithecus pentelici* has been documented in Greece and Italy and is dated to the late Miocene/early Pliocene (Turolian) (Rook 1999; Tsoukal and Bartsiokas 2008). The dental characteristics of Mesopithecus have demonstrated affinities with Asian colobines, particularly the extant so-called odd-nosed monkeys, *Rhinopithecus* and *Nasalis* (Pan, Groves, and Oxnard 2004). *Dolichopithecus* is another genus of colobine monkey known from sites in Spain, France, and Central Europe, with *Dolicopithecus balcanicus* from Bulgaria dated to around 3.5 mya (Spassov and Geraads 2007).

Pliocene cercopithecine monkeys are known from several species belonging to the genus *Macaca*. The progenitor of the genus *Macaca*, to which the macaques belong, was present in North Africa approximately 7–8 million years ago, and the ancestor of the Asian macaques evolved about 5.5 mya and had migrated eastward to northern India by 3 million years ago (Smith, Mcdonough, and George 2006). By 2 million years ago, *Macaca* diverged into three phylogenetic groups and dispersed into Southeast Asia in three waves: the silenus group,[5] the sinica-arctoides group, and the fascicularis group (Abegg and Thierry 2002; Smith, Mcdonough, and George 2006). By the time the macaques arrived in Southeast Asia, many were recognizably related to extant species. One example, *Macaca sylvanus* cf. *prisca,* is dated to about 3.2 mya in Italy (Rook, Mottura, and Gentili 2004). In more recent times, *Macaca nemestrina*, *Hylobates syndactylus* (siamang), and *Pongo pygmaeus* have been found in Java in flowstone deposits 128,000–118,000 years ago (Westaway et al. 2007). Longtail macaques (*Macaca fascicularis*) expanded throughout insular Southeast Asia (Sundaland) including the Philippines by at least 40,000 years ago (Smith, Mcdonough, and George 2006).

Homo erectus and Other Hominins

The position of *Homo erectus* in modern human evolution has undergone major revision over the several decades. *Homo erectus* had long been believed to represent an intermediary stage in human evolution in Africa whereby *Homo habilis* gave rise to *Homo erectus*, which gave rise to *Homo sapiens* (see review by Antón 2003). According to this view, *Homo erectus* migrated into Asia, where it thrived in the Middle Pleistocene (~150,000–800,000 ya), and was eventually replaced by modern *Homo sapiens.* It is now known that *Homo erectus* was in Eurasia much earlier and much later than the Middle Pleistocene. The date of the *Homo erectus* radiation into Eurasia is now placed at closer to 2 million years ago instead of 1 million, and it appears to have been present in some parts of Southeast Asia within the last 50,000 years. Asian *Homo erectus* may be actually be older than African *Homo erectus* (= *Homo ergaster*), meaning that the two could be sister taxa. Here, instead of African *Homo erectus* migrating into Asia, a hominin form earlier than both *H. ergaster* and *H. erectus* may have been responsible for populating Asia.

Antón, Leonard, and Robertson (2002) have suggested that the initial dispersion of *Homo erectus* into Eurasia may have been precipitated by a complex of biogeographic factors. In their view, the increase in brain and body size for *Homo erectus* compared to earlier hominins likely increased their dependence on animal fats and proteins. Ecological changes led to increased aridity and expansion of grasslands, while continental changes opened pathways between Africa and Eurasia with extensive faunal exchanges. *Homo erectus* may have migrated with the dispersing herbivores.

In Java, *Homo erectus* fossils have been dated as early as 1.8 mya, at the Late Pliocene/Early Pleistocene boundary (Larick et al. 2001; Swisher et al. 1994). This would place them at an earlier date than the oldest known *Homo erectus/ergaster* in Africa, raising the possibility that *Homo erectus* (sensu stricto) evolved in Asia (Kaifu et al. 2005). Although there has been some skepticism for this very early date (for example, de Vos and Sondaar 1994), its feasibility has been supported by additional early finds elsewhere outside Africa. A number of recently identified fossils from Dmanisi in the Georgia Caucasus date as early as 1.77 mya (Lordkipanidze et al. 2006). In addition, the site of Longgupo Cave in China may date as early as 1.9 mya (Huang et al. 1995). Numerous fossils from Zhoukoudian cave in China provide a continuous record of *Homo erectus* from around 800,000–400,000 years ago (Boaz et al. 2004; Shen et al. 2009; Wu, Schepartz, and Norton 2010). *Homo erectus* has also been identified at the Hexian site in southern China at about 412,000 years ago (Wu, Schepartz, and Liu 2006). In Turkey, *Homo erectus* has been dated to around 510,000–490,000 years ago.

The Middle Pleistocene terminal date for *Homo erectus* in Asia has been extended into the Late Pleistocene. In Java, *Homo erectus* has been dated to 53,000–27,000 years ago (Swisher et al. 1996) and to 70,000–40,000 years ago (Yokoyama et al. 2008). These dates would make *Homo erectus* broadly contemporaneous with *Homo sapiens* in the region. Although *Homo sapiens* does not appear in the fossil record in Java until after 10,000 years ago, *Homo sapiens* may have been present in Malaysia by 40,000 years ago (Swisher et al. 1996). The Niah Cave site in Sarawak, Borneo, has recently been redated, indicating *Homo sapiens* occupation at 34,000–46,000 years ago.

In addition, a new small-brained, small-bodied hominin species has been identified from Flores, Indonesia, designated as *Homo floresiensis* (Brown et al. 2004). It has dated from 95,000–74,000 to 12,000 years ago, but its taxonomic status is unclear (Morwood et al. 2005). Cranial analyses have yielded quite different interpretations, including affinities with *Homo erectus* (Brown et al. 2004; Falk et al. 2005; Gordon, Nevell, and Wood 2008) unattributable to any known *Homo* species (Argue et al. 2006; Morwood et al. 2005) and abnormal microencephalic modern *Homo sapiens* (Jacob et al. 2006; Martin et al. 2006a, 2006b). If *Homo floresiensis* proves to be a late relict population derived from *Homo erectus*, it would extend the time frame of the species in Indonesia to 12,000 years ago. Regardless of the taxonomic status of *Homo floresiensis*, for the purposes here, it does provide evidence of hominins in island Southeast Asia until the near the end of the Late Pleistocene, with evidence of *Homo sapiens* in the fossil record shortly thereafter. In sum, island Southeast Asia can be characterized as having a continuous hominin occupation from nearly 2 million years ago to the present.

In conclusion, the fossil record suggests a third possibility for the origin of vivax-type malarias. The data suggest a continuous record of hominoid primates in Asia from the time of the Miocene apes to the present. Although vivax-type malaria is most diversified today in macaques and other Southeast Asian monkeys, the origin could very well have been in any one of the hominoids entering Asia. In other words, a host-switch from Asian monkeys to modern humans or from modern humans to Asian monkeys is not the only possibility. Ancient hominins or other hominoids could have harbored vivax-type malaria. Although the hominoid record is characterized by many extinctions and replacements, vivax-type malaria could have been sustained in hominoids through host-switches among species—meaning that although individual species may have disappeared, vivax-type malaria could have been maintained among hominoids over time. Colobine and cercopithecine monkeys left Africa and began entering Eurasia around 10 million years ago. Miocene apes or later hominins could have laterally transferred vivax-type malaria to Asian wild primates at any point from 10 million years ago to recent times. However, such a scenario is only one possibility. The strength of Carter's version of Pleistocene refuge theory is that it offers an explanation for the existence of *P. gonderi* as a relict form existing in a West Africa refugia; Pleistocene refuge theory may also help explain African ape malaria distribution. However, Carter's model of a Trans-Pacific entry of vivax-type malaria into the New World is more problematic. The origin of vivax-type malaria in the New World is the topic of the next chapter.

As a final note, I mention two exciting new discoveries in hominin evolution. Green and associates (2010) recently provided a draft sequence of the Neanderthal genome after extracting DNA from 40,000 year old bones. A very long-standing question in paleoanthropology has been whether the Neanderthals interbred with modern humans (for example, Evans et al. 2006; Trinkaus et al. 2003) or were replaced by them (for instance, Bailey, Weaver, and Hublin 2009; Currat and Excoffier 2004). Green and members of the Neanderthal Genome Project found that approximately 2.5% of the genome of non-African humans comes from Neanderthals. Thus, some interbreeding did indeed take place between modern humans and Neanderthals. In addition, a new hominin from Denisova, Siberia, was recently discovered dating to between 30,000 to 48,000 years ago (Krause et al. 2010). The Denisovans appear to be related to Neanderthals but differ enough to warrant designation as a separate species. They apparently interbred with the ancestors of Papuan New Guinea peoples, whose genomes are estimated to be about

5% Denisovan. Although these findings are somewhat peripheral to the vivax question, they are important to recognize as a warning about drawing hard conclusions from the available data in paleoanthropology and molecular biology. The best we can do is to provide tentative syntheses and interpretations.

Notes

1. Steiper et al. (2005) cite the dissertation of N. D. Wolfe (1999) for the molecular phylogeny linking *P. pitheci* and *P. silvaticum* to the vivax group.
2. The deletion was also found in an unidentified parasite in *Mandrillus leucophaeus*, designated *Plasmodium* sp. DAJ-2004.
3. The time of the arrival of the platyrrhines in South America at around 30 million years ago is based both on the fossil record in South American and morphological similarities and differences between Old World primates and New World primates. However, there is debate regarding the route taken by the platyrrhine ancestors that led to their arrival in South America. See Cormier (2003) for a review of literature and key controversies.
4. It should be noted that these genetic relationships are still far from settled. A phylogenetic analysis conducted by Mu and associates (2005) did not determine clear distinctions between Old World and New World forms of *P. vivax* or a closer connection of *P. simium* to Old World *P. vivax*. Nevertheless, Mu and associates (2005) found support for an Asian origin of *P. vivax*-like malarias, which were transmitted to New World humans and monkeys through a series of host transfers.
5. The ancestor to the barbary macaque (*Macaca sylvanus*) is also included in the silenus group, but all extant *M. sylvanus* are in Europe and Africa.

Chapter 5

Migration: Malaria in the New World

Little doubt exists that the colonization of the New World included the transfer of malarial parasites from the Old World. Most researchers view human (and primate) malaria in the New World as a virgin soil epidemic. However, some debate has existed as to whether or not some form of primate malaria may have also existed in the New World before colonization. The endemic New World malaria hypothesis holds that one or more forms of malaria were present in the New World in Pre-Columbian times. The hypothesis takes three forms. One is that malaria was present in the first Americans who carried the disease from Asia, across Beringia, into the Americas. The second form suggests that Post-Beringian and Pre-Columbian Southeast Asian seafarers introduced malaria into South America. In these two scenarios, New World monkeys are believed to have anthroponotically acquired malaria from humans. The third form suggests that Neotropical monkey *P. brasilianum* and *P. simium* were laterally transferred from monkeys to humans giving rise to *P. malariae* and *P. vivax*, which spread from South America into Old World populations.

This chapter explores the evidence for the spread of New World malarias.

New World Malaria

Although some debate exists regarding the origin of *P. vivax* and *P. malariae* in the New World, *P. falciparum* is more straightforward. Africans were likely the primary source of falciparum malaria in the New World. In the 15th century, malaria was endemic in all the major European colonizing nations. Europeans likely harbored and introduced vivax malaria, although they may have also been an additional source of falciparum malaria. As described in Chapter 2, genetic polymorphisms that offer some resistance to falciparum malaria are found not only in

Africa but also in Asia and the Mediterranean, including portions of southern Europe. In addition, falciparum malaria may have been on the rise in European populations owing to colonial enterprises in Africa and the institutionalization of slavery in the Old World as well as in the Americas. The enslaved central and west African peoples were unlikely to have harbored vivax malaria, owing to RBC Duffy negativity. The origin of the malariae form is less clear. *Plasmodium* malariae is present at much lower levels in Africa than falciparum malaria is today; it may have also been present historically in Europe. Because *P. malariae* is a "benign" malaria, its symptoms are often difficult to distinguish from the symptoms of *P. vivax* in the historical records before diagnostic smears were available.

Several lines of evidence suggest that falciparum malaria was introduced to the New World with colonization. One, New World indigenous peoples lack high rates of prevalence of the polymorphisms that provide resistance to falciparum malaria, such as sickle-cell trait (for example, Allison 2009; Weatherall and Clegg 2001), α and β-thalassemia (for instance, May et al. 2007; Vichinsky et al. 2005; Weatherall and Clegg 2001), and glucose-6-phosphate-dehydrogenase (G6PD) (for example, Capellini and Fiorelli 2008; Nkhoma et al. 2009; Weatherall and Clegg 2001). In addition, although the malariae-type and vivax-type malarias are today endemic in New World monkey populations, falciparum-type malaria is not. However, very recently, *P. falciparum* was identified in two species of howler monkeys (*Alouatta*) in the wild (Duarte et al. 2008). Although the presence of *P. falciparum* in wild howlers appears to be a rare isolated instance of anthroponotic infection in a few monkeys, New World monkeys are susceptible to *P. falciparum* under experimental conditions. That suggests that New World monkeys could potentially serve as a reservoir for *P. falciparum*. But there is no evidence that New World monkeys are natural hosts to any form of falciparum-type malaria. In addition, the high burden of *P. falciparum* malaria in Africa suggests that African slaves (and/or infected Europeans) are the source of the *P. falciparum* found today in indigenous New World peoples.

Although there is no serious debate that *P. falciparum* was introduced into the New World at colonization, some debate exists as to whether other forms of malaria may have existed in human or wild monkeys in the New World before colonization. Historically, the Pre-Columbian malaria hypothesis has been argued based on cultural evidence. For example, evidence of pictures of mosquitoes on pottery from New Mexico and translations of several Mayan words have been argued to be suggestive of Pre-Columbian malaria (Wood 1975); however, this type of cultural evidence is quite thin. Obviously, the presence of mosquitoes does not mean that malaria was present.

One long-standing line of evidence that has supported a hypothesis of Pre-Columbian malaria comes from the South American medicinal plant cinchona, which has antimalarial activity. The antimalarial drug quinine derives from this plant. Honigsbaum and Willcox (2004; see also Cormier 2010) have reviewed the debate surrounding whether cinchona[1] was used as an antimalarial by indigenous South Americans. Spanish chroniclers of the 17th century describe cinchona in southern Ecuador (then, the Loja region of the Spanish Vice Royalty of Peru). Bark from the plant was first brought to Europe by Jesuit priests. It is assumed that they learned of the medicinal properties of the plant from Quechua-speaking indigenous South Americans. Today, Ecuadorian healers use the plant for a variety of purposes, including as a treatment for diarrhea, to induce labor, and as a dentifrice. What is unclear is whether or not cinchona was specifically used as an antimalarial before European contact. Honigsbaum and Willcox (2004) suggest, because of the absence of a record of cinchona in Incan hieroglyphics or archaeological remains, that it was adopted more recently as a traditional remedy. In addition, no record has been found of cinchona being used indigenously as a febrifuge.

However, the absence of a record of cinchona in the archaeological record of the Inca does not mean it was not used elsewhere. Today, there are over 10 million speakers of Quechua in South America (Hornberger 1998). Differences exist in regional dialects, cultural traditions, and ecological contexts. Of all the Quecha speakers, the Inca were perhaps the least likely to have developed a treatment for malaria, owing to the relative difficulty of plasmodia being able to reproduce in mosquitoes in the cold, dry, high-altitude climate of the Andes. If cinchona was used as a malarial remedy, it would have been more likely to have been used among lowland tropical cultures; however, plant remains are much more difficult to recover from archaeological sites in the humid tropical forest. An additional possibility is that cinchona began to be used indigenously as an antimalarial only after contact. If it had been a traditional medicinal plant, its usage could have been adopted to combat a new disease threat. If malaria had been introduced, cinchona certainly would have had become established by the time of the 17th-century Spanish chroniclers.

Another key reason to suspect that malaria is a recent introduction to the New World is its epidemiological patterns among nonhuman species. In the Old World, plasmodia-causing malarias are found in all major groups of terrestrial vertebrates (Wood 1975). Such cross-species diversification suggests a long evolutionary history. In the New World, malaria has a much more restricted species distribution, occurring primarily in humans and monkeys. The exceptions are that in North America,

Caribbean *Anolis* lizards can be infected with *Plasmodium floridense* and *P. azurophilum* (Schall and Vogt 1993). Also, *Plasmodium odocoilei* is found in the North American white-tailed deer (*Odocoileus virginianus*). However, we do not find the type of diversification of malaria across mammals in the New World that is found in the Old World. The presence of these plasmodia in a few isolated Old World species (apart from primates) may very well be recent introductions. But more to the point, because the deer and lizard plasmodia are such distant relatives to the primate malarias, they could not have possibly given rise to the primate malarias. A related point is that New World monkey plasmodia have not diversified, which suggests a much shallower time depth. Although malaria occurs in over 30 species of New World monkeys, they represent only two plasmodia species—a stark contrast with Southeast Asian primate malaria.

By way of analogy, clearer ethnohistorical evidence exists for the introduction of another arthropod-borne infection, yellow fever, during the colonial era. Wide agreement exists that both the *Flavivirus* disease agent and the *Aedes aegypti* mosquito vector of yellow fever were introduced from Africa to the Americas during the slave trade (for example, Dobyns 1993; Kiple and Higgins 1992; Wirsing 1985). Neotropical monkeys presumably acquired the disease from humans. In addition, African populations and Old World primate populations also demonstrate immunological resistance to yellow fever lacking among Native Americans and Neotropical monkeys (Kiple and Higgins 1992).

The evidence seems scant in support of any primate malaria plasmodia having a long evolutionary history in the New World. However, a third possibility exists—namely, that primate malaria did not originate in the New World and that malariae or vivax type did not arrive with the European colonization but were introduced by Pre-Columbian human groups. Two potential scenarios exist: one, that the first Native Americans carried malaria into the New World when migrating west from northern Asia to northern North America. The second is Carter's (2003) hypothesis, discussed in the previous chapter, that malaria entered the New World in Pre-Columbian times from Southeast Asia by a seafaring route.

Beringia: By Land or by Sea

It is generally agreed that a North Asian Siberian population migrated by land or sea toward the Bering Strait land bridge and dispersed from that area into the Americas no later than 16,500 years ago (Goebel, Waters, and O'Rourke 2008). This hypothesis was originally based on archaeological data, and recent genetic evidence confirms that contemporary

Native Americans ancestry is in Asia (Karafet, Zegura, and Hammer 2006; Merriwether 2006). According to the early form of the Beringia model, the first Americans were believed to have crossed the Bering Strait bridge by land, hunting big game, and spread within a few centuries from the northern tip of North America to the southern end of South America (Haynes 2002).

One of the problems with the land-route hypothesis is that it involves an extremely rapid dispersal of Native Americans across two continents. In addition, the archaeological evidence demonstrates the earliest North and American sites occurring roughly at the same time. In South America, the Monte Verde site in southern Chile has been dated to around 14,600 years ago (Dillehay et al. 2008). In North America, the dating of Meadowcroft Rockshelter in Pennsylvania remains controversial, but the Page-Ladson site in Florida and the Paisley Cave site in Oregon provide evidence of occupation by around 14,600 years ago (Goebel, Waters, and O'Rourke 2008). The problem with these North and South American sites is not only that there is significant distance between them but also that intervening sites of similar date have not been found that would support trekking across the continents. Dillehay and associates (2008) have found evidence of marine algae at the Monte Verde site, likely used for food and medicine, and believe this could mean that a coastal route was taken to settle the area. The two hypotheses are not mutually exclusive, however. Humans may have entered the New World both overland across Beringia and along a sea route, off the coast of Beringia and down the west coast of the Americas.

The possibility of malaria entering the New World with the first Americans is usually discounted owing to the "cold filter" hypothesis. Anopheline mosquitoes were not present in the cold Arctic passage, giving the plasmodia no vector stage (Kelton 2007). The cold-filter hypothesis is also bolstered by archaeological evidence that the population that migrated into the Americas had adapted to the climate of the Siberian Arctic by 32,000 years ago. It is also known that *P. vivax* and *P. malariae* plasmodia will not develop in anopheline mosquitoes whose body temperatures fall below 15°C (59°F) (Ramenofsky 1993).

Taken together, these data do seem to exclude the possibility of malaria entering the New World with the first Americans. However, there are several reasons why the hypothesis should not be immediately discounted. One, *P. vivax* is a relapsing malaria. As discussed in Chapter 2, *P. vivax* undergoes a hypnozoite stage where it remains dormant. Although the period of dormancy is typically a year, or possibly two, the upper limit of the period of dormancy is not known. For example, historical records from Finland between 1750 and 1850 showed that *P. vivax* remained dormant from between two and nine years (Huldén,

Huldén, and Heliövaara 2008). The longest period of dormancy recorded is 30 years (Trampuz et al. 2003; White 2003). Also, as discussed in Chapter 2, the mechanism that appears to trigger plasmodia from coming out of dormancy is the biting of mosquitoes. In contemporary populations where *P. vivax* is present, long dormancies would not likely occur, because mosquitoes would be available. However, it is possible that the period of dormancy is indefinite in a human host.

Although *P. vivax*-type malaria could conceivably exist in a dormant stage in a human host throughout his or her lifetime, the bigger issue is whether anopheline mosquitoes and their more temperature-sensitive plasmodia could survive. Even for *P. falciparum,* which is commonly thought of as a tropical disease, epidemics of *P. falciparum* malaria were reported throughout the 19th century in Siberia, as far north as the Arctic seaport of Arkhangelsk (61° 30′N) (Reiter 2008). Another important consideration is that the temperature limitations for plasmodial reproduction are often thought of in terms of ambient air. In the United States after colonization (Humphreys 2001) as well as 18th-century Finland (Huldén, Huldén, and Heliövaara 2008), mosquitoes overwintered in human habitations, in barns and in households. Although plasmodia would have a difficult time producing in the natural environment in cold climates (or cold seasons), they would be able to do so in anthropogenic habitats.

That being said, it still seems that the possibility of malaria accompanying the first Americans into the New World is rather remote. The ecological barriers would have been formidable. Although small populations of plasmodia-infected anopheline mosquitoes could have conceivably accompanied humans across the Arctic, the likelihood of the mosquitoes keeping warm enough in primitive structures for the plasmodia to reproduce seems unlikely. Even given the potentially long hypnozoite stage of *P. vivax* and a potentially more rapid coastal entry into South America for anthroponotic transmission to New World monkeys, a human to monkey lateral transfer seems very improbable. Even the 19th-century Arctic epidemics likely do not represent sustained infections, but the episodic spread of malarial infections northward. Migrating Asian populations would have been increasing the distance between them and areas where malaria may have been endemic, making the possibility of transmission less likely as time passed.

THE TRANS-PACIFIC HYPOTHESIS

The possibility that human populations other than Native Americans made incursions into the Americas before the European colonization not only has been the subject of scientific inquiry but has also captured the popular imagination. It is now known that at least two populations entered the Americas by boat after the initial peopling of the Americas and before 1942. From the Atlantic, archaeological evidence has demonstrated that between 800 and 1000 C.E., the Norse established short-lived settlements in Newfoundland and Maine (McGovern 1990). Although the Norse presumably had contact with Native Americans, their temporary incursions into the extreme north would not be possible sources for South American Neotropical monkey malaria. However, it has recently been established that Pacific-traveling peoples also entered the Americas.

The possibility of early Polynesian and South American contact was popularized following Thor Heyerdahl's publication of the book *Kon Tiki* (1950). It was an immediate best-seller, and a motion picture adaptation won the Academy Award in 1951. The basis of his book was a Polynesian legend that the islands were once inhabited by long-eared and short-eared races of people. As the legend goes, a war broke out between the two groups, and the short-eared people killed all the long-eared people. Heyerdahl proposed that the long-eared race originated in South America and traveled by boat to Polynesia. To prove it, he built a raft and sailed it from Peru to the Tuamotu Islands in French Polynesia. Although the scientific community essentially scoffed at the hypothesis that an extinct race of long-eared people from South America were once on Polynesia, Heyerdahl had in fact demonstrated that the length of the Pacific Ocean could be crossed with primitive boat technology. Fifty years later, his hypothesis proved to be partially correct, but the Trans-Pacific route may have been established first in the other direction. Recent evidence suggests that Polynesians entered South America, leaving behind the domesticated chicken.

The domesticated chicken (*Gallus gallus*) in South America had long been assumed to have arrived with European colonists. However, recent DNA and archaeological evidence has demonstrated that domesticated chickens were present in South America before the arrival of Europeans and originated in Polynesia (Storey et al. 2007). Radio-carbon analysis of a chicken bone from the El Arenal-1 site on the Arauco Peninsula in Chile provides a date of 1321–1407 C.E., approximately 100 years before the European colonization. The bone also produced an identical sequence to two chicken bones at two prehistoric archaeological sites in Polynesia.

The apparent Polynesian introduction of the domesticated chicken to South America has the potential to explain the mystery of how sweet potatoes (*Ipomoea batatas*) arrived in Polynesia; although it is a South American domesticate, it is today an important crop in Polynesia. The oldest archaeological evidence of sweet potatoes in Polynesia derives from carbonized remains of tubers in Mangaia Island in Central

Polynesia at 1000–1100 C.E. (Green 2005; Kirch 2000). Possible early explanations were that sweet potatoes were dispersed to Polynesia by birds or accidental casting by vessel (O'Brien 1972). Likewise, the linguistic similarity between the Quechuan term "kumur" and the Easter Islands term "kumara" was seen as a late Quechuan borrowing from the Spanish, who adopted the term from Polynesia (Montenegro, Avis, and Weaver 2007; O'Brien 1972). However, current evidence that Polynesians traveled to South America and introduced the domesticated chicken also raises the possibility that they may have also returned to Polynesia carrying the sweet potato. Although such a trip would have been a tremendous undertaking, ethnographically it makes more sense that Polynesians, renowned for making distant trips by sea, would have made the voyage both ways than that South Americans would independently have undertaken a voyage to Polynesia. Nonetheless, if a Trans-Pacific exchange of domesticates occurred in Pre-Columbian times, the possibility exists that diseases may have also been transferred.

When malarial geneticists identified the close relationship between *P. vivax* and *P. simium*, that suggested a lateral transfer, in some direction. One possibility that was proposed to account for the relationship was that vivax-type malaria originated in the Neotropics and was transferred as a zoonosis from New World monkeys to New World humans (Ayala, Escalante, and Rich 1999; Escalante et al. 1998). It should be stressed that the geneticists who proposed this hypothesis did not view it as a very likely scenario but rather considered an alternative hypothesis that could potentially explain the close relationship between *P. vivax* and *P. simium*. If such a transfer occurred, it could have been through a *Kon Tiki* route in Pre-Columbian times or as recently as 500 years ago with the European colonization. The latter scenario seems unlikely, because if vivax-type malaria originated in the Neotropics, it would have had to spread rapidly worldwide to human populations and diversified in Asian primates in only the last few hundred years, with colonization. This scenario is also at odds with the historical literature. As discussed in Chapter 1, the evidence suggests that a malaria form consistent with *P. vivax* symptoms was present in Europe, Asia, and the Mediterranean dating back several millennia.

An anthroponotic transfer of vivax-type malaria from humans to monkeys in the New World is the more likely scenario. One study that refuted the hypothesis of a Neotropical monkey origin of *P. vivax* also raised the possibility of a Pre-Columbian Trans-Pacific transfer. As mentioned in Chapter 4, a study by Li and colleagues (2001) found derived differences between Old World and New World forms of *P. vivax*. More surprisingly, they found that Neotropical monkey *P. simium* clustered with the Old World *P. vivax* variant rather than the New World *P. vivax* variant. In other words, the study suggested that there is a closer genetic relationship between Asian human *P. vivax* and New World *P. vivax* than either has to the form of *P. vivax* occurring in contemporary New World peoples. They proposed that human South American vivax be redesignated as a new subspecies, *P. vivax collensi*, or even a new species, *P. collensi*.

From an anthropological perspective, it is difficult to explain the historical events that could account for the relationship between Neotropical monkey vivax and Asian human vivax. Because human *P. vivax* is closely related to multiple forms of vivax-type malaria in Southeast Asia, a lateral transfer likely occurred between humans or their ancestors or wild primates or their ancestors in that region. Regardless of whether the vivax-type ancestor ultimately originated in Africa or Asia, Southeast Asia appears to be the area where a host-switch occurred. It seems that the close relationship between human *P. vivax* and *P. simium* can be explained logically only by a human-Neotropical monkey host-switch, because Asian primates and Neotropical primates have been separated geographically for at least 30 million years. Thus, the genetic relationship between Asian and Neotropical wild primate vivax-type malarias can be explained only by a human intermediary.

As also discussed in Chapter 4, Carter (2003)—drawing on both the work of Li and colleagues (2001) and Pleistocene refugee theory—proposed a hypothesis for the entry of *P. vivax* into the New World by way of a Pre-Columbian Trans-Pacific transfer. Here, he proposed that Neotropical monkeys harbor a Neanderthal variant of *P. vivax*. In his model, human *P. vivax* was isolated in Mediterranean and Southeast Asian refuge areas. The Mediterranean variant was present in Neanderthals and transferred to modern humans as they migrated out of Africa. It then became established in modern European populations and was spread to the New World during colonization. The Southeast Asian variant was spread to the New World by seafaring humans who brought the disease to South America, where it entered the human population and was transferred to monkeys. Carter speculates that the Asian form of *P. vivax* introduced into South America eventually disappeared from the original human carriers because of unknown environmental or cultural behaviors that prevented its perpetuation. The proposed *P. vivax collensi* was introduced by Europeans and represents a form that was eradicated from that continent in the mid-20th century.

Possible support for Carter's model is found in the behavior of *Anopheles cruzii*, a mosquito vector of *P. simium*. Unlike many of the anopheline vectors of human malaria, *An. cruzii* is a sylvatic species, spending up to 99% of its time in the upper canopy (Lourenço-de-Olveira and Deane 1995; see also Deane et al. 1970). Two of the species that carry *P. simium*, the woolly spider monkey (*Brachyteles arachnoides*) and the brown howler monkey (*Alouatta guariba*), also spend most of their time in the emergent levels of the forest or the upper canopy[2] (Kinzey 1997). Thus once *P. simium* was transmitted, the habitats of the monkeys and their plasmodium vector may have provided a natural barrier, making it difficult to transfer *P. simium* from monkeys to humans in the wild.

However, potential ecological barriers in the vertical forest niches of monkeys do not explain why *P. simium* has failed to spread more widely among South American monkeys. It is limited to three species, all of which occur in the Atlantic Coastal forest. Members of the genus *Alouatta* are found throughout the tropical forests in South America, spanning the continent from coast to coast with the exception of the highland Andes (Emmons and Feer 1997). More to the point, a major problem with the Trans-Pacific hypothesis is that the distribution of *P. simium* is on the wrong side of the continent. To account for this distribution, a hypothetical human population would have to have sustained malaria long enough to trek at least a thousand miles in order to transmit the disease to Atlantic Coast howlers and woolly spider monkeys. If it had been sustained long enough in humans to make such a long journey to the Atlantic Coast, it would have unlikely disappeared without infecting any other human populations along the way. At the very least, *P. simium* would be expected to have been transmitted to one of the many intervening monkey populations, particularly *Alouatta* species. If an ecological barrier was present, it seems more likely to be found in the contemporary conditions of the highly developed and fragmented Atlantic forest. In addition, not all geneticists agree that *P. simium* clusters more closely with human Asian *P. vivax* than human South American *P. vivax*. A phylogenetic analysis by Mu and associates (2005) did not find a closer connection of *P. simium* to Old World than to New World *P. vivax*.

Regardless of whether the hypothesis of a twice-introduced *P. vivax* is eventually supported, the problem still remains of explaining the disjunct distribution of *P. simium* in only a few Atlantic Coast monkey species. Rather than a Trans-Pacific Pre-Columbian entry, a simpler explanation would lie with a much more recent introduction. I propose that vivax-type malaria may have been introduced by 19th-century Chinese agricultural laborers into Brazil (see also Cormier 2010). During this time, tea was an exotic commodity highly desired by the Brazilian elite. In 1814 a colony of 200 Chinese workers was brought to Rio de Janeiro to cultivate tea for the Royal Botanical Garden (Marchant 1961). Over the course of a century, approximately 3,000 laborers were brought from the Portuguese colony of Macao in southern China to work in Rio de Janeiro (Conrad 1975; Lesser 1999; Marchant 1961).

If two variants of *P. vivax* did indeed enter the New World, a Chinese origin of *P. vivax* would seem a more viable alternative hypothesis to a Trans-Pacific route; it is a far more parsimonious explanation than Southeast Asians traveling by boat to the New World. Even given the recent evidence of boat travel from Polynesia in Remote Oceania, human malaria is not endemic even today in the Pacific east of Vanuatu and the Solomon Islands (WHO 2010). For *P. vivax* to have been brought directly from Southeast Asia where *P. vivax* is present would have entailed nearly doubling the traveling distance from Remote Oceania in primitive watercraft in Pre-Columbian times. Geographically, the location of the migrant Chinese migrant laborers matches the distribution of the Neotropical primates that harbor *P. simium* in the Atlantic forest. The agricultural activities of the Chinese laborers involving the felling of trees for tea cultivation may have increased their likelihood of coming into contact with the sylvatic *An. cruzii*. If this is the case, then the presence of vivax-type malaria in Neotropical primates may ultimately be linked with the desire of Brazilian colonists to drink tea.

Even if the hypothesis of *P. vivax* being twice introduced is not supported, the current distribution of *P. simium* is likely due to ecological variables. Whether the original human host was Asian or European, environmental disturbance is likely to have circumscribed *P. simium* in monkeys. Africans who were brought to the New World in the slave trade are an unlikely source, due to RBC Duffy negativity. The coastal Atlantic forest was a site of early colonial enterprises and was rapidly deforested. The highly disturbed and fragmented nature of the Atlantic forest likely explains why *P. simium* has not spread farther westward in Neotropical primates. In addition, there is some ethnohistorical evidence that supports the hypothesis that all malaria was introduced after colonization. Joralemon (1982) describes the absence of reports of malaria among early explorers who visited Amazonian groups in riverine habitats, where malaria is found today. According to Joralemon, it is not documented until after 1650 in the northeast and eastern coasts of South America. He argues that this is compelling evidence for a late introduction.

In sum, the evidence seems to point to the introduction of *P. vivax* malaria into the New World at the time of colonization. New World primates acquired the disease anthroponotically from humans, and it is now designated as *P. simium*. The relationship between *P. brasilianum* and *P. malariae* has not yet been explored. Because *P. malariae* is relatively benign and occurs at lower rates than either *P. falciparum* or *P. vivax*, it receives less attention in the literature. But as will be seen in the next chapter, the rates of *P. brasilianum* may be underestimated in South America. Like the relationship between *P. simium* and *P. vivax*, many geneticists have argued that the relationship is so close between *P. brasilianum* and *P. malariae* that they should be considered indistinguishable conspecifics (for example, Escalante, Barrio, and Ayala 1995; Ewald 1983; Fandeur et al. 2000; Lal et al. 1988; LeClerc et al. 2004; Rich and Ayala 2000). Thus many of the same hypotheses regarding host-switching equally apply to *P. malariae/P. brasilianum*. However, the possibility of *P. brasilianum* existing in primates before colonization is much less likely due to its relationship to African primate malaria plasmodia. Early on it was suspected that the chimpanzee *P. rodhaini* was actually human *P. malariae* (Coatney et al. 1971). In addition, *P. malariae* has been identified in chimpanzees and may be a natural parasite (Duval et al. 2010; Hayakawa et al. 2009; Kaiser et al. 2010). The presence of *P. brasilianum*, which may be conspecific with African human *P. malariae* and chimpanzee *P. rodhaini*, bolsters the argument for all forms of malaria arriving in the New World at the time of colonization.

The Broader Context

In the bigger picture, if malaria indeed entered the Americas as a virgin soil epidemic, its significance should be treated in contest with the devastating effects of all introduced diseases to North and South American indigenous populations. Introduced Old World diseases literally decimated indigenous New World populations from European early explorations through the early colonial era (for example, Cook 1998; Denevan 1992b; Dobyns 1983, 1993). Malaria has generally been considered to have been one of these introduced diseases (see also Dunn 1965; McNeill 1976; Wirsing 1985). As is well known, many of the diseases affecting New World human populations were presumably zoonoses, such as small pox, chicken pox, measles, and influenza (for instance, Crosby 1994; Haggett 1994).

Old World domestication of animals allowed the opportunity for zoonotic transfer of animal diseases to human populations, and dense population settlements allowed continuous spread of the diseases from human host to human host. Over time in the Old World, co-evolution between the pathogen and the host appears to have led to both host increased resistance and attenuation in virulence of many zoonotic Old World diseases. Indigenous New World peoples lacked such selection of population resistance to these diseases. But the malarias present a more complicated picture than other introduced diseases such as small pox and the influenzas because they cannot be passed directly from human to human; they require a mosquito vector. As previously described, these then involve triadic parasite-vector-host adaptations, and increased virulence may be selected for. Moreover, since a vector is involved, malaria cannot be understood without taking into account the ecological conditions that affect mosquito proliferation.

In terms of settlement patterns, some Pre-Columbian Andean and Mesoamerican populations (and likely Amazonian chiefdoms as well) were densely populated enough to allow for human chain-to-chain infection of zoonoses. And, indeed, when the zoonotic diseases were introduced, they spread rapidly through these populations with devastating effects (for example, Alden and Miller 1987; McNeill 1989: 215–218). However, New World Pre-Columbian groups lacked the kinds of animal domestication that fostered the zoonotic spread of disease seen in the Old World. Animal domestication in South America was limited to the Andean highlands, where three animals were domesticated: the alpaca, the llama, and the guinea pig. Although the dog may not have been originally domesticated in the New World (Savolainen et al. 2002), this animal domesticate was present among in the New World before European colonization (Schwartz 1997: 34–36). According to McNeill (1989: 210–211), in the Andean highlands, the llama and the alpaca lived in dispersed populations rather than in herds large enough to sustain chains of infectious disease. The domesticated guinea pig lived inside households, and evidence suggests that it did serve as both a host and a reservoir of at least one parasitic vector affecting humans, Chagas disease (Coimbra 1988: 89–90). Thus there is a qualitative difference in at least the degree of animal domestication in South America; animal domestication was limited in the highlands and absent among lowland Amazonian peoples.

Strictly speaking, malaria does not conform well to the category of zoonosis, because its history in the Old World seems to implicate potential cross-infections at the level of anthropoid primates rather than a one-way transfer from nonhumans to humans. Nonetheless, it appears to have entered the New World in much the same way of other Old World zoonoses, introducing new disease forms to populations previously unexposed. In the case of malaria, its proliferation is not a simple case of infectious transfer from human to human. Because vectors are involved, the rise of malaria in North and South America must be understood in terms of landscape transformations that fostered the proliferation of malaria. Nonetheless, much of our historical understanding of malaria among human and wild primates derives from experimental research, which is discussed in the next chapter.

Notes

1. Cinchona is a member of the Rubiaceae family and is not a single plant, but includes from 23 to 36 different species, depending on the botanical classification (Honigsbaum and Willcox 2004).
2. *Alouatta caraya* is more variable in the canopy strata utilized.

Chapter 6

Rhesus Factor: Experimental Studies in Wild-Primates

The aim of this chapter is to explore how we know what we know about the primate malarias through experimental research. The early history of experimental studies involving attempts to infect human and wild primates with malaria is discussed. Broadly stated, the goal of this experimental research was to develop effective primate models for the study of human malaria in order to better understand the disease with the hope of developing effective antimalarial treatments. In the process, attempts have been made to introduce novel forms of malaria to wild-primate hosts, which has indirectly provided much information about the susceptibility of wild primates to various forms of malaria.

The use of wild primates in biomedical research for malaria did not originally derive from an understanding of the close relationship of the primate malarias but rather from historical factors that led to increased reliance on Rhesus monkeys as the gold standard in biomedical research, beginning with polio studies in the 1940s and solidifying with the thalidomide studies and Harry Harlow's psychological experiments in the late 1950s and early 1960s. It eventually became apparent than neither Rhesus monkeys, nor any of the other Old World monkeys tested were useful in the study of the most virulent form of malaria, *P. falciparum,* because infections were difficult to establish. As previously discussed, a mutation not present in Old World monkeys seems to have arisen in New World monkeys, making them vulnerable to human *P. falciparum* infection. However, the mutation appears to have been selected for under conditions unrelated to malaria itself.

When it was discovered that Rhesus monkeys could not serve as models for falciparum malaria, New World monkeys began to be utilized in malarial research. Although any number of species of New World monkey could have been used for research, owl monkeys (*Aotus* spp.)

have become the wild primate of choice, and, to a lesser extent, squirrel monkeys (*Saimiri* spp.). Owl monkeys have generically become the primate model for both studies of falciparum malaria as well as vivax malaria. However, numerous studies have involved experimental cross-infections of multiple primate species with novel malarias; these have involved infecting wild primates with human malarias, infecting humans with wild-primate malarias, and infecting wild primates with novel wild-primate malarias. The responses vary considerably, including natural resistance in some species, virulent reactions, and within-species individual differences. This chapter also evaluates these findings in light of the hypothesis of differential susceptibility to vivax-type malaria driving the speciation and geographic distribution of primate malarias, with special attention to Southeast Asia.

The Historical Context of Wild-Primate Experimental Studies

Attempts have been made to induce a number of forms of malaria in a number of primate species, but the vast majority of the literature on malaria in primates has involved Rhesus macaques (*Macaca mulatta*) and owl monkeys (*Aotus* spp.). The choice of these primates has been as much because of their availability as it has been their suitability as animal models for malaria research. Thus the context of primate research has to be understood as also taking place within a political arena where science and ethics have often conflicted. By far, the most widely used primate in biomedical and psychological research has been the Rhesus macaque. Three lines of research have been important historically in characterizing the role of Rhesus macaques in medical experimentation: the polio vaccine, thalidomide birth defects, and the Harry Harlow social deprivation studies. Polio created the demand for large-scale imports of wild primates, thalidomide established the imperative of drug testing in primate species, and the Harlow experiments, arguably, marked the beginning of ethics in animal research.

Poliomyelitis is a viral disease that invades the nervous system, and in approximately 1 in 200 cases, it leads to irreversible paralysis (WHO 2008a). The causative agent of polio was identified in 1908 by Karl Landsteiner and Erwin Popper (Blum 1994; Kantha 1995; Kristensson 1999). They took spinal fluid from a child who had died from polio and injected it into the spinal cord of rabbits, guinea pigs, mice, and monkeys (two individuals of each type). None of the animals were affected except for the monkeys. One, a Rhesus macaque (*Macaca mulatta*), developed paralysis after 17 days; the other, a Hamadryas baboon (*Papio hamadryas*), died after 8 days. Landsteiner (with A. S. Wiener) subsequently went on to discover the Rhesus blood group system in the 1930s, including the so-called Rh Factor.

In the 1940s and 1950s Jonas Salk and Albert Sabin had been independently working on a polio vaccine (Blum 1994; Sabin 1985). The need for a vaccine became critical after epidemic polio swept the United States in the early 1950s, leaving an estimated 22,000 persons paralyzed each year between 1950 and 1954 (Trevelyan, Smallman-Raynor, and Cliff 2005). When both the vaccines were initially developed, not only were monkeys needed to test the vaccines but also their kidney cells were used to create the vaccine itself (Blum 1994). One monkey could be used to produce 65 vaccines; however, if the ultimate goal was to vaccinate all of humanity, massive numbers of monkeys would be required. By 1955 the U.S. government had signed an agreement with India for the export of Rhesus macaques and was importing 10,000 of them each month, just for the development of the polio vaccine alone (Loftin 1955). That same year, Salk's first clinical trials led to 400,000 children being vaccinated in the United States, and by 1960, 15 million Soviet citizens had been given Sabin's vaccine (Blume and Geesink 2000). It has been estimated that by the late 1950s, the United States was importing 200,000 Rhesus macaques per year, and overall at least 1 million monkeys died in polio research alone (Blum 1994).

In *Monkey Wars* (1994), Deborah Blum argues that although polio provided the success story, thalidomide provided the disaster that established primates as essential for biomedical research. Thalidomide was developed by Ciba, a Swiss pharmaceutical company in 1953, and by 1960 it was being marketed as a sedative and antinausea medication in 40 countries (Rajkumar 2004). The drug was tested extensively in laboratory rodents and was deemed safe enough to give to pregnant women to relieve morning sickness (Knobloch, Shaughnessy, and Rüther 2007; Rajkumar 2004). The consequences were horrific, creating severe birth defects resulting in missing or flipper-like limbs, eventually effecting over 10,000 infants, who came to be known as the "thalidomide babies" (Botting 2002). In the United States, the tragedy was averted; Kelsey Frances blocked its approval by the FDA, because he believed that it had not been adequately tested (Rajkumar 2004). Subsequent testing on primates revealed the same limb deformities that were seen in humans, but not in rodents (Knobloch, Shaughnessy, and Rüther 2007). As a consequence, it came to be viewed as essential to use primate models in medical research, because distantly related species, such as mice and rodents, may not respond in similar ways to humans.

A counter-current emerged almost simultaneously regarding the use of wild primates in experimental research, largely because of the psychological experiments of Harry Harlow with Rhesus monkeys. Harlow conducted studies in both maternal deprivation and total isolation with infant Rhesus monkeys, sometimes placing infants in complete isolation for up to 24 months (Blum 1994). When removed from isolation, Harlow described aggression and extreme psychological disturbance, including, "autistic self-clutching and rocking" (Harlow, Dodsworth, and Harlow 1965:92). Other experiments involved maternal deprivation through providing the monkey with a substitute mother made of wire or even one that would blast air and shake vigorously to knock the infant off, or an "iron maiden" that would stab the infant if he or she attempted to cling (Blum 1994). One of Harlow's doctoral students, Gene Sackett, has argued that Harry Harlow's experiments were the genesis of the modern animal-rights movement (Blum 2002). Thalidomide and Harlow perhaps epitomize the ethical dilemma that continues today: because nonhuman primates are close relatives to humans, they provide the best biomedical models; yet because nonhuman primates, particularly post-Harlow, are recognized as our socially complex cousins, use of nonhuman primates in medical research is intensely scrutinized for ethical considerations.

The question of animal ethics was not limited to the Western world. In the mid-50s, almost immediately after a U.S.-India import for Rhesus monkeys was signed, shipping was temporarily banned in 1955 after several hundred Rhesus monkeys were accidentally suffocated in a layover on a flight en route to the United States (Anonymous 1961). Indian officials charged the United States with cruelty and neglect, and the incident outraged Hindus, who view the monkey as sacred (Oppenheimer 1977; Small 1994; Wheatley 1999; Wolfe 1991, 1992, 2002). The massive imports of monkeys also led to radical population declines in the wild in India. In one study conducted between 1960 and 1980, the population of Rhesus monkeys in agricultural areas decreased from 60% to total disappearance in some areas (Southwick, Siddiqi, and Oppenheimer 1983). The decline was attributed not only to massive export of Rhesus monkeys before the ban in 1978 but also to complications of human population growth and deforestation.

By the early 1960s researchers were already looking for alternatives to the Rhesus monkeys, primarily because of concerns about their continued availability. Although New World monkeys are more distantly related to humans than Old World monkeys, they were considered to offer other advantages in being less expensive to import to U.S. laboratories and being easier to maintain because of their smaller size (Marley 1963). Although I have not seen this specifically mentioned in the literature, I suspect that owl monkeys are particularly easy to handle because they are the only nocturnal monkeys. From my own research in primate pet-keeping practices in Amazonia, owl monkeys are extremely active at night but quite docile and drowsy during the day. Among the Guajá, pet monkeys travel with their owners by clinging to the tops of their heads. On several occasions, I witnessed owl monkeys falling off Guajá heads because they had fallen asleep. During the day-time work hours of human laboratory researchers, sleepy owl monkeys should be relatively easy to manage. Rhesus monkeys, in contrast, are notoriously difficult to handle. According to Blum (1994), the street drug PCP (phencyclidine hydrochloride) was originally developed in 1956 as a sedative to control aggressiveness in laboratory Rhesus macaques.

Broad Brush Differences in Asian, African, and South American Wild-Primate Malarias

Experimental studies on wild-primate cross-infections with novel forms of malaria were largely conducted in the 1960s and 1970s. The goal was to find appropriate primate models that could serve to provide insight into the course of malarial infections in humans. These largely involved studies attempting to infect wild primates with human malarias. For the reasons described, eventually owl monkeys became the primate of choice, because they were susceptible to and relatively tolerant of the two major forms of malaria than afflict humans: *P. falciparum* and *P. vivax* (see Appendix 1); that is, owl monkeys have been considered to be good models for both of these forms of human malarias because it is easy to infect them and because their reaction is not so severe that they suffer a high fatality rate, enabling the disease course to be studied.

In the course of evaluating potential primate models, experimental studies tested not only different primate species as models for infection with human malaria but also the susceptibility of various primate species to novel wild-primate malarias. The rationale was driven by both logistical factors and early erroneous typologies based on morphology and life-history features. For the former, concerns about the availability of Rhesus monkeys for continued research led researchers to seek alternative primate species that could be used as models. For the latter, early typologies based on similarities in gross morphology of the plasmodia species and life histories (quartan, tertian, and so on) suggested relationships among plasmodia that have today been disproven, such as *P. simiovale* being closely related to *P. ovale* (it is actually a vivax-type malaria) and *P. fragile* being related to human falciparum malaria (it is also a vivax-type malaria) (see Coatney et al. 1971). While these studies were not designed to systematically test the response to malaria in all primate hosts, they have been instructive in understanding how different plasmodia affect different primates. The successes and failures of these studies are a rich source of data regarding potential differences in genetic adaptations of primates to malaria. The experimental studies that provide the most insight about potential biological resistance are those involving attempts to induce malaria into a novel host that has not been experimentally altered and is also closely related to the natural host.

A few words of caution are necessary for interpreting the cross-infection studies. For some of the malaria studies, it is not only the laboratory setting itself that provides the experimental context. Some of the studies have involved splenectomized primates—splenectomy alters the immune response.[1] In others, malaria is first cycled through a primate host that tolerates the parasite before one attempts to produce an infection in a secondary novel host. For reasons that are not well understood, such cycling can alter virulence and/or susceptibility (Escalante et al. 1998; McCutchan 2008; Palatnik and Rowe 1984; also see neurosyphilis studies, following).

In addition, we should not mistake the scientific *evidence* for real-world actualities. For example, we know far more about malaria in Rhesus macaques (*Macaca mulatta*) than we do, say, in the stump-tailed macaques (*Macaca arctoides*). That does not necessarily mean that the expression of malaria in Rhesus monkeys ultimately has greater relevance for understanding malaria in primates than does the expression of malaria in the stump-tailed macaques. We have literally, thousands of biomedical studies that use Rhesus macaques as a model and only a handful that use stump-tailed macaques. Cultural, political, and historical factors, rather than scientific rationale, have been the basis for using Rhesus macaques as a model. Certainly, the species more closely related to humans than the Rhesus macaque in Asia, gibbons and orangutans, could potentially provide better models for understanding human malaria—and better still our African hominid cousins, bonobos, chimpanzees, and gorillas. Again, scientific studies were not necessarily based on studies of the *best* species to give insight into malaria but on the *available* species for scientific testing.

In addition to human *P. vivax*, 10 other vivax-type malarias have been identified in wild primates, eight of which occur in Southeast Asia. Although it remains unclear if that region is the origin of vivax-type malaria, it is a least a center of endemicity and species expansion. As described in Chapter 4, five additional plasmodia of Southeast Asian malarias of gibbons and orangutans likely belong to the vivax group (see complete reference list for vivax-type malarias in Appendix 1). Most of the species diversity among the wild-primate malarias occurs within Asian cercopithecine and colobine monkeys but is particularly pronounced in macaques species. One of the gibbon malarias that infect *Hylobates moloch* has been determined to belong to the vivax group. Other gibbon plasmodia that are suspected to belong to the vivax group are *P. eyelesi, P. jeffreyi,* and *P. youngi*, and the orangutan plasmodia suspected to belong to the vivax group are *P. pithici* and *P. silvaticum*. Thus, including humans, we are likely dealing with at least 14 different members of the vivax group occurring in Southeast Asia.

In contrast to Southeast Asia, only one definitive vivax member has been identified in wild primates in Africa. *Plasmodium gonderi* occurs in a circumscribed area in three species of mangabeys (*Cercocebus agilis, C. atys,* and *Lophocebus aterrimus*) and the drill (*Mandrillus leucophaeus*). The two other potential African monkey vivax-type malarias are *P. georgesi* (in *C. agilis* and *L. albigena*) and *P. petersi* (*L. albigena*). Thus it does not appear that malaria is wildly distributed in African monkeys. However, a number of Africa guenon and baboon species have been demonstrated to be susceptible to cross-infection with other wild-primate vivax-type plasmodia under experimental conditions. These are listed below (see Appendix 1 for references).

Experimental African Monkey Hosts of Novel Malaria Parasites

Cercocebus atys	*P. knowlesi*
Cercopithecus aethiops	*P. cynomolgi, P. gonderi, P. knowlesi*
Cercopithecus cephus	*P. knowlesi*
Cercopithecus mitis	*P. inui*
Cercopithecus mona	*P. knowlesi*
Papio anubis	*P. fieldi, P. gonderi, P. inui, P. knowlesi*
Papio cynocephalus	*P. gonderi, P. knowlesi*
Papio papio	*P. cynomolgi, P. knowlesi*
Papio ursinus	*P. gonderi*

In South America, distribution of malaria in wild monkeys bears some similarity to both the patterns found in Southeast Asian and African monkeys. *Plasmodium brasilianum* is extremely widespread among New World monkeys occurring naturally in at least 35 species of monkeys and spanning geographically from Central America to southern Brazil. Given that the evidence suggests that *P. brasilianum* is an anthroponotic infection of monkeys, it likely dispersed rapidly throughout the New World primate population within the last 500 years. On the one hand, the case of *P. brasilianum* provides a cautionary tale for presuming that the widespread distribution of vivax-type malaria in Southeast Asia necessarily means that it is a center of origin. On the other hand, two key differences exist in comparing the Southeast Asian and South American cases. One is that *P. brasilianum* belongs to the malariae group, rather than the vivax group. In humans, *P. malariae* (as well as *P. ovale*) has not been well studied owing to its relatively benign symptomology in comparison to falciparum and because it occurs at much lower levels of frequency in human populations than *P. vivax* does. (But as an aside, to be discussed later, the frequency of *P. malariae* may be underestimated in indigenous Amazonians.) Nonetheless, *P. brasilianum* is not well understood in terms of plasmodial life history, host susceptibility variables, or perhaps even more important, ecological factors related to potential vectors. The second key difference is that although *P. brasilianum* spread rapidly and widely among South American primates, it is the only species of the malariae group affecting Neotropical primates. In Southeast Asia, considerable *plasmodial species* diversification exists. Since multiple vivax-type species of plasmodia are represented in Southeast Asia, this does suggest that some degree of host-plasmodia cospeciation occurred, in addition to potential lateral host-switching. To put it another way, if we did a thought experiment and assumed that the original primates that colonized the New World harbored a malaria plasmodia, we would expect that over the course of 30 million years of evolutionary history, the basal plasmodial ancestor would have diversified so that multiple species of plasmodia would be found in the New World.

The African and South American situations are similar in that vivax-type malaria has failed to expand beyond a narrowly circumscribed zone. This is true in both geographic regions despite the existence of species that likely could support vivax-type malaria. For the baboons (*Papio* spp.), ecological conditions may provide an explanation, since many live in savannah habitats that may not be as suitable to support plasmodial vectors (Fleagle 1998). However, vervet monkeys (*Cercopithecus aethiops*) who are susceptible to *P. gonderi* and other vivax-type malarias are widely distributed in tropical forests east to west, from coast to coast (Fleagle 1998). And, as previously described, vivax-type malaria has not been documented in mangabeys (*Cercocebus* spp.) living in tropical forest areas much beyond the coastal area.

South America presents a similar situation, with *P. simium* extant in two howler monkey species (*Alouatta)* in the Atlantic forest but not in other howlers that are distributed from southern Mexico to northern Argentina (Emmons and Feer 1997). Under experimental conditions, *P. simium* infects two additional howler species, as well as spider monkeys (*Ateles*), owl monkeys (*Aotus*), capuchins (*Cebus*), woolly monkeys (*Lagothrix*), squirrel monkeys (*Saimiri*) and two species of callitrichids (*Callithrix* and *Saguinus*).

EXPERIMENTAL CROSS-INFECTIONS

Although the distribution of vivax-type malarias in Southeast Asian and South American monkeys suggests that ecological factors are likely at play, experimental cross-infections have demonstrated differences in the way that individual species react to individual plasmodia. The responses of wild primates to novel forms of malaria have varied considerably. The most informative studies regarding the relationships among primates and plasmodia are those of the vivax group, in which closely related plasmodia species produce very different reactions in closely related primate hosts. In some cases, it has not been possible to produce symptomatic infection, even by direct inoculation with sporozoites in splenectomized hosts. In other cases, the reaction to a novel plasmodium species has been so severe that it has been imminently life threatening. In addition, for some types of plasmodia in some primate species, individual differences exist with some members of the same species developing only mild symptoms and others becoming critically ill.

The seven species of vivax-type malaria that occur in Southeast Asian monkeys vary in their degree of host-specificity. Two of the Asian monkey plasmodia may be host-specific. The Toque macaque (*Macaca sinica*) is the only monkey species that has been identified as a natural host of *P. simiovale,* and the long-tailed macaque (*Macaca fascicularis*) appears to be the natural host of *P. coatneyi* (although found at low densities in *M. nemestrina*). Other plasmodia of Asian monkeys infect multiple species, such as *P. inui,* which has 11 different Asian monkeys as natural hosts, including the Toque macaque and the long-tailed macaque. It infects both macaques (cercopithecines) and langurs (colobines) and so is not even specific at the genus level, affecting monkeys in different subfamilies. However, *P. inui* is the only known plasmodia to infect the Assamese macaque (*Macaca assamensis*). In addition, many of the Asian monkeys are hosts to more than one species of plasmodium, such as the long-tailed macaque, who is host to *P. coatneyi, P. cynomolgi, P. inui,* and *P. knowlesi.*

Although it is clear that some of the plasmodia infect more than one species of Asian monkey and some Asian monkeys are hosts to more than one species of plasmodia, it is more difficult to make definitive statements about host-specificity. The biggest obstacle is that we have very little data on the prevalence of malaria in wild primates. We know even less about their mosquito vectors. In the case of the macaques, there are approximately 20 Asian species, but malaria-causing plasmodia have been isolated in only 8 of them. This does not necessarily mean that malaria does not occur in the other species; rather, there is no evidence at this point. And, as stressed, some species, such as the Rhesus monkey, have been far better studied than others.

Few systematic studies have been done to assess the extent of malaria in Asian monkeys, with the exception of a large-scale study done in the 1960s in Kuala Lumpur (Slater 2005, and see following). In some cases, the isolation of plasmodia in monkeys can be misleading if they are taken from either zoo or laboratory settings. For example, Coatney and associates (1971) describe the isolation of *P. cynomolgi* in a stump-tailed macaque (*Macaca arctoides*) housed in a zoo, which appears to be the only documented case. Since *P. cynomolgi* has been documented in eight other Asian monkey species, it is quite possible that stump-tailed macaques are not a natural host but acquired the infection from other monkeys either in the zoo environment or possibly through being housed with other monkeys when being transported for export.

Ideally, if we had more data on the prevalence of malaria-causing plasmodia among the Asian monkeys, we might possibly construct a distribution map of the extent of a given plasmodial species, either within a given monkey species or geographically across hosts, as well as a map of known vectors. Without a doubt, local ecological conditions play a role in the distribution of malaria in wild primates as they do in human populations; we just do not have enough information at this point to make valid statements about geographic range of wild-primate plasmodia. In many cases, a particular plasmodium is isolated in a primate species only after the primates undergo a medical examination in a research facility. Here, the specific environmental context may be lost, and it might not be accurate to generalize from the malarial case to the potential range based on the distribution of the host. For example, Rhesus macaques (*Macaca mulatta*) have been described as having the broadest geographic distribution of any primate apart from humans (Southwick et al. 1996). They are found in India, China, Afghanistan, and Southeast Asia. However, that does not mean that the two plasmodial species they harbor (*P. cynomolgi* and *P. inui*) have the same broad geographic distribution as those in Rhesus monkeys. By the same token, we do not find *P. falciparum* in every ecological area that humans inhabit.

One example that might put the problem into perspective is the overlapping distribution of a number of monkeys in the border area of Myanmar and China. Five different malaria-causing plasmodia have been identified in five different monkey species in this area. However, some of the plasmodia involved infect more than one host, and some of the monkeys involved can be infected with more than one type of plasmodia. Also, although five primate species are present in the area, they are at the southern edge of the Assamese macaque range, the northern edge of the range of the pig-tailed macaque (*Macaca nemestrina*), and in the middle of the range of Rhesus macaques. Let's say that a given plasmodium, X, were found in the laboratory with all three species (which happens to be the case with *P. inui*). From that information alone, it cannot be determined if malaria X is limited to their shared geographic range or extends broadly across the ranges of all three. Particularly for the studies of the 1960s and 1970s, when many of the primate malarias were isolated, geographic context is not often available. It is also possible that plasmodium X is associated with a specific environmental feature, such as riverine habitats that may occur disjunctly across a broad region.

Despite these caveats, experimental studies have provided information about the differences in symptomology of hosts that are introduced with what are believed to be novel forms of malaria. These experimental studies do provide evidence that closely related primates can differ significantly in their immune response to closely related forms of malaria. Although the implications of these differences can be interpreted in several ways, they do at least suggest that host resistance may be a factor in understanding the distribution of the plasmodia among the Southeast Asian monkeys.

One example of variation in host resistance among closely related primate species is the response to *P. fragile,* which occurs naturally in the bonnet macaque (*Macaca radiata*) and the Toque macaque (*Macaca sinica*). In a study involving a blood-induced infection of *P. fragile* in bonnet macaques, Rhesus macaques, and Hanuman langurs (*Presbytis entellus*), the course of the disease differed for these three species (Puri and Dutta 1994). Bonnet macaques, a natural host, experienced only mild symptoms, whereas the Rhesus monkeys and Hanuman langurs became critically ill. It proved fatal to half the Rhesus monkeys and all the Hanuman langurs in the study. Similar virulent responses were found when *P. fragile* was introduced into more distantly related hosts, owl monkeys (Collins et al. 1974b). Other plasmodia have been identified as occurring in Rhesus macaques (*P. cynomolgi* and *P. inui*) and in the Hanuman langur (*P. cynomolgi*). Given the mild symptoms in the bonnet macaque and the severe symptoms in Rhesus monkeys and Hanuman langurs (as well as two species of New World monkeys), we might guess that bonnet macaques have developed a degree of biological resistance to *P. fragile*. As previously described, the level of virulence of a particular plasmodial species in a given host does not necessarily indicate the amount of time that a host and pathogen have coevolved. But in a situation where the same plasmodium species is benign in one species and lethal in others, it could suggest biological adaptation on the part of the host.

Another example is *P. coatneyi,* which occurs naturally in the long-tailed macaque. It produces critical disease in Japanese macaques (*Macaca fuscata*) and Rhesus monkeys. In one study comparing *P. coatneyi* in long-tailed macaques and Japanese macaques, the long-tailed macaques experienced only mild symptoms, whereas 100% of the Japanese macaques died after experiencing severe T-cell lymphocytopenia (Matsumoto et al. 2000). Japanese macaques have also demonstrated symptoms similar to those seen in human falciparum malaria with the sequestration of parasitized red blood cells in cerebral microvessels (Kawai et al. 1993; Kawai et al. 2006; Sugiyama et al. 2004). Rhesus macaques have exhibited similar symptoms to the Japanese macaques in cerebral vessels (Aikawa et al. 1992; Nakano et al. 1996; Sein et al. 1993) and have also been shown to produce proinflammatory cytokines in the brain, which is also characteristic of human cerebral malaria (Tongren et al. 2000). In addition, Rhesus monkeys infected with *P. coatneyi* also exhibit symptoms similar to human placental malaria with intrauterine growth retardation and low-birth-weight infants (Davison et al. 1998; Davison et al. 2000). In one case study, a Rhesus macaque infected with *P. coatneyi* developed DIC (disseminated intravascular coagulation), which essentially involves systemic blood coagulation that depletes the platelets and results in massive hemorrhage (Moreno et al. 2007).

Again, at first glance, these studies would seem to provide classic examples of host-pathogen co-evolutionary processes. When hosts and pathogens are in long association, natural selection *may* operate over time, resulting in a more resistant host population and an attenuated pathogen population. In these cases, at the level of the host, individuals in a given population who are naturally more resistant than others to a given pathogen have greater reproductive success than do those who are less resistant. At the level of the pathogen, virulence may be selected against. Pathogens that kill their hosts may have less reproductive success than milder forms. As previously described, examples of such co-evolutionary processes can be seen with introduced Old World diseases that literally decimated indigenous peoples in North and South America who had not developed resistance during the colonial era (for example, Cook 1998; Denevan 1992b; Dobyns 1983, 1993). However, as elaborated in Chapter 2, the relationships are more complicated with vector-borne diseases, where virulence in hosts may not be attenuated over time. Nonetheless, with those caveats, a plasmodium species that creates a mild reaction in one primate host species and a virulent reaction in another primate species may suggest that the first dyad has undergone co-adaptation, whereas the second may represent a recent introduction into a novel host. However, even cases of benign reaction to malaria, such as the bonnet macaque reaction to *P. fragile*, must also be interpreted with caution. For example, owl monkeys and squirrel monkeys are novel hosts for all the Old World vivax malarias they have been exposed to experimentally. However, both have varied considerably in their reaction to infection with novel malarias. Owl monkeys experience very low levels of parasitemia when infected with *P. hylobati,* a gibbon malaria (Collins et al. 1974a), but Asian-monkey *P. knowlesi* is fatal to them (Collins et al. 1981a). While *P. coatneyi* has been demonstrated to produce severe disease in novel macaque hosts, attempts to induce *P. coatneyi* in owl monkeys (Collins et al. 1981a) and squirrel monkeys (Sullivan et al. 2005) have not been successful. Similarly, attempts to infect owl monkeys and squirrel monkeys with *P. gonderi* have also failed (Sullivan et al. 2002). Owl monkeys are questionably susceptible to *P. fieldi* (Sullivan et al. 1998). Experiments with *P. fragile* have produced differing results at the subspecies level. *Aotus trivirgatus trivirgatus* monkeys developed fulminating infections with *P. fragile,* but an infection could not be induced at all in the same study with *A. trivirgatus griseimembra* monkeys (Collins et al. 1974b).

Similarly, differences have been seen in *Macaca fascicularis* populations from different locales (Migot-Nabias et al. 1999). *P. coatneyi* proved fatal to long-tailed macaques from Mauritius[2] but not to those from the Philippines. Differences have also been seen among individuals of from the same population. Olive baboons (*Papio anubis*) experimentally infected with *P. knowlesi* experienced two disparate outcomes (Ozwara et al. 2003). They either developed acute and severe disease accompanied by cerebral involvement or they developed a milder chronic form of the disease. Thus although one might be led to believe that the benign symptoms of *P. fragile* in the bonnet macaque suggest co-evolution, that could not possibly be the explanation for the benign symptoms of gibbon *P. hylobati* in the owl monkey. Although owl monkeys clearly have biological resistance to *P. hylobati*, their resistance had to have evolved for reasons unrelated to exposure to that plasmodium and, in fact, likely unrelated to any malaria, since it appears that New World monkey malaria is a recent introduction.

These results taken as a whole suggest that several types of evolutionary processes could have occurred among the Southeast Asian primates in terms of host-pathogen relations, all of which could have occurred to varying degrees. To begin with, it can be taken as a given that there was a process of vivax-plasmodium speciation in Southeast Asia primates. Including gibbon vivax malaria, there are at least eight different vivax-type species that are all very closely related to one another occurring in the same general geographic region. One explanation is that natural differences in host resistance led to speciation of plasmodia within the host. Here, an ancestral vivax plasmodium entered the Southeast Asian primate population and encountered species with differing levels of immunity. Plasmodium speciation could have involved adaptation to the biological environment of different primate hosts. The differences in natural biological resistance to the differing forms of malaria in owl monkeys lend some support to this view. For example, if owl monkeys were consistently exposed to *P. fieldi* over time, that plasmodium might evolve so that it could establish itself more readily in that host; similarly, with long exposure to *P. knowlesi*, that plasmodium might attenuate to become established in that host. One study that provides some support for plasmodium evolution involves differences in trophozoite preferences by species and by plasmodia. Warren, Skinner, and Guinn (1966) found that *P. knowlesi* selectively invaded reticulocytes in long-tailed macaques (*M. fascicularis*) but showed no preference for cell type in Rhesus macaques (*M. mulatta*). *P. cynomolgi* preferentially invaded the reticulocytes in both macaque species. For *P. coatneyi* and *P. inui*, they invade mature erythrocytes in both macaque species. Although the significance of this preference is not known, it does demonstrate that different forms of plasmodia react differently to the red blood cells of different hosts.

It is also possible that vivax-type malaria itself drove primate Asian-monkey speciation. Wheatley (1980) first proposed the hypothesis that macaque speciation was a response to vivax-type malaria. Using the examples of Rhesus (*M. mulatta*) and long-tailed macaques (*M. fascicularis*), he argued that malaria could be the mechanism that delimited their ranges and that the acquisition of genetic immunity of the long-tailed macaque facilitated its divergence from the ancestral macaque. The Rhesus macaque and the long-tailed macaque both have wide distributions but are allopatric (their ranges essentially do not overlap). In general, the Rhesus macaque is in the northern areas of Southeast Asia, and the long-tailed macaque is farther south. In Wheatley's scenario, malaria existed in Southeast Asia before a radiation of ancestral macaques to the area. Given that long-tailed macaques tolerate several types of vivax malaria that create severe illness in Rhesus monkeys, he proposed that Rhesus and long-tailed macaques diverged owing to differences in biological resistance in the ancestral macaque population. Here, long-tailed macaques evolved from those among the ancestral population that were better able to tolerate malaria and move southward into heavier malarial zones. One line of evidence supporting Wheatley's view are the natural differences seen in olive baboons exposed to *P. knowlesi*, with some developing acute infection and others developing mild chronic infection. If such differences existed in an ancestral macaque population, they possibly could have driven speciation.

These are not necessarily either/or scenarios, because co-evolutionary processes could have involved mechanisms of both host selection and pathogen selection; two other possibilities may have come into play. One is host-switching. With malaria, it cannot be assumed that because a plasmodium is well-established in a primate population that it has necessarily had a long history of co-evolution. Again, malaria in the New World primates provides an example. *Plasmodium brasilianum* apparently entered the primate populations after colonization as an anthroponotic form of human *P. malariae.* In a relatively short period of time, it has become widely established both in terms of species and geographic distribution, found in over 30 New World monkey species from Central America to Southern Brazil. The level of virulence is not necessarily a reliable indicator of the time depth of malaria in a given population. Falciparum malaria has been established long enough in human populations for a number of biological adaptations to evolve that provide some degree of resistance. Yet, falciparum malaria remains generally, a rather virulent disease in humans, particularly for children.

The second possibility is mosquito-mediated selection. Co-evolution does not involve just the host-parasite dyad but also the host-parasite-vector triad. The mosquito is the most fragile part of the malaria chain, because of its short life span and the very narrow time frame that is available for the plasmodia to complete the sexual stage of its reproduction in the mosquito. Ultimately, the most important co-evolutionary processes may be occurring at the level of the mosquito and the plasmodia. It may very well be that human and nonprimate host resistance is not nearly as important as has been assumed for plasmodia survival. As discussed in Chapter 2, as long as a host is at least infectible, it may not matter if the course of the illness is long or short, acute or chronic. For the plasmodia, the essential need is that the primate host live long enough for the plasmodia to complete its life cycle within the host, only a matter of weeks.

The next chapter addresses experimental studies with humans. Although not discussed in this chapter, one additional reason why cross-infections were attempted in various monkey species was the concern that humans could be infected with forms of wild-primate malaria. Most of these studies took place in the 1960s and 1970s. In addition, at one time, malaria was believed to be an effective treatment for late-stage syphilis, and monkey malarias were used in these experiments. These early studies typically involved vulnerable human populations and would be considered unethical by today's standards. Nonetheless, they are part of the history of malarial research and have provided much information about human susceptibility to wild-primate malarias.

Notes

1. The spleen has important role in the disease process of malaria, being a key site for removal of parasitized erythrocytes, generation of immunity, and production of new erythrocytes (Engwerda, Beattie, and Amante 2005).
2. The long-tailed macaques were introduced to Mauritius in relatively recent times, which fact may have some relationship to differences in their reaction.

Chapter 7

Ethics: Human Experimentation

This chapter reviews early human experiments involving malaria. Many involved the testing of antimalarial drugs but were also driven by concerns of the potential for wild-primate malarias to be transferred to humans zoonotically. The purpose of this chapter on the human malaria trials is not to review all human experimentation but to discuss some of the ethical issues in early human experimentation and what was learned about human susceptibility to wild-primate malarias. Such information is valuable to reexamine today, because ecological changes are bringing human and wild primates into contact, making the possibility of lateral transfer of disease more likely.

A watershed event that led to attempts to cross-infect primate species with novel forms of malaria was the accidental transmission of wild primate *Plasmodium cynomolgi* to malaria researchers in 1960 in the United States. Although endemic malaria had been effectively eradiated in the United States by mid-century, cases of malaria in U.S. citizens returning from travel abroad raised the specter of malaria becoming reestablished on U.S. soil. The World War II-era Stateville prison malaria studies are well documented, but less attention has been given to the Atlanta Penitentiary studies of the 1960s.

The Stateville malaria project involved infecting prisoners with human malaria and testing antimalarial drugs. The Atlanta Penitentiary studies involved intentional exposure of inmates with wild-primate malarias to determine whether infections could be established. Before 1960 it was not believed to be possible for humans to acquire a nonhuman primate form of malaria, until a researcher was accidentally infected with *P. cynomolgi* (common in macaques). Publications, particularly in the *American Journal of Tropical Medicine and Hygiene,* revealed that over a 10-year period, several hundred prisoners were inoculated with various strains of wild-primate malaria. Wild-primate malaria was also used for a short time as a treatment for neurosyphilis.

This chapter also includes various isolated cases of humans naturally acquiring wild-primate malarias. The recent outbreaks of macaque *P. knowlesi* in Indonesia are addressed in the final chapter.

INFECTIONS OF PRISONERS WITH WILD-PRIMATE MALARIAS

The Thirty-Second Annual Charles Franklin Craig Lecture delivered before the American Society of Tropical Medicine and Hygiene, Philadelphia, Pennsylvania, November 1, 1967:

> With the advent of the second accidental human infection with monkey malaria in 1960, blood was drawn from the patient and inoculated into two inmate volunteers at the U.S. Penitentiary, Atlanta Georgia. Those initial inoculations inaugurated a program of investigation dealing with simian malaria in man that I am happy to say, is still in progress. Up to mid-September of this year 204 volunteers have been exposed to infections with simian malaria... we and other malariologists salute these unselfish volunteers. Such participation is unique because this is the only country in the world where such voluntary participation is permitted, encouraged, and actually sanctioned by law (Coatney 1968:152).

In the 1960s and 1970s a number of studies were conducted in the prison population of the Atlanta Penitentiary to determine if it was possible to infect humans with various strains of wild-monkey malaria. According to Coatney (1968), by the end of the 1960s over 200 prisoners had been involved in these studies. During that time such experimentation was sanctioned by both the government and the medical profession and reflected a long history of these types of experimental studies on prisoners and other vulnerable populations through the mid-1970s. Inmates had been used in medical experimentation for some time, and many of the earlier prison studies are particularly troubling. For example, between 1918 and 1922, prisoners in San Quentin California received human and animal testicle transplants, and prisoners at Rankin Farm Prison in Mississippi were fed a niacin-poor diet in order to induce pellagra (Hornblum 1997).

Ironically, the Nuremberg war crimes tribunals (1945–1949) greatly expanded the use of prison populations in medical experimentation. One of the largest prison studies was initiated at Stateville Penitentiary in 1944, which involved over 500 prisoners exposed to *P. falciparum* malaria in order to test chemotherapy drugs (Alving et al. 1948; Hornblum 1997). In Hornblum's (1997) review, he describes both the role of the Nuremberg trials and the extent of the types of prison studies that followed, summarized in the next paragraph.

During the Nuremberg trials, the defense made the argument that there was no difference between Nazi concentration camp studies and U.S. prison studies, giving particular attention to malaria research at Stateville. The argument for the prosecution was that the key difference was that the American prisoners were volunteers, and thus these studies were to be contrasted as examples of high standards of ethical medical practice. This led to a great expansion of these prison studies, including exposure to sleeping sickness (African trypanosomias), gonorrhea, gas gangrene, dengue, hepatitis, syphilis, amoebic dysentery, influenza, flash burns, blood with live cancer cells, dioxin, and even athlete's foot. Public attention to these experiments came about in the early 1970s, following the exposure of the Tuskegee Institute Syphilis studies. Before Nuremberg, prison studies involving malaria primarily involved testing of antimalarial drugs. Table 7.1 provides a list of the available studies on these types of trials in the United States.

The Stateville prison studies on antimalarial drugs were well publicized and today are a staple of college introductory psychology texts in discussions of ethics in research. Malaria research at the U.S. Penitentiary in Atlanta, Georgia, is less well known. In fact, Coatney's Atlanta Penitentiary studies were used as a model for the experiments done at Stateville (Alving et al. 1948). From the published studies, the Atlanta Penitentiary studies involved more prisoners than the Stateville experiments, with over 700 involved in the testing of antimalarials and over 200 subjected to attempts to infect with wild-primate malaria.

The context for the interest in exposing humans to wild-primate malarias has been described by the NIH (National Institutes of Health) historian, Leo Slater (2005). In 1960, after a researcher, Don Eyles, who had been studying primate models of malaria was accidentally infected with *Plasmodium cynomolgi* from a mosquito bite, the cross-infection generated serious concern. Previously, it had not been thought possible to transmit monkey malaria to humans through a vector. According to Slater, such transmission posed a new threat for the malaria eradication programs in place and also posed a new health threat owing to the potential for new forms of malaria to be introduced to U.S. populations. This situation led to the establishment of a large-scale study in Kuala Lumpur to determine the extent of simian malaria in humans. At the end of the study, it was determined that wild-primate malaria did not pose a serious health threat to human populations.

In the United States, Eyles's accidental infection also marked the initiation of approximately 10 years of experimentation on Atlanta inmates, attempting to establish various types of simian malaria in humans. This study involved experiments using both mosquito vectors and parasitized blood. The vector approach involved studies attempting to transmit malaria by mosquito bite from an infected monkey to an uninfected

Table 7.1 Antimalarial Drug Tests among Atlanta, Jessup, and Statesville Prisoners

Location		Number of Prisoners	Reference
Atlanta[1]	*P. vivax*	36	Coatney et al. 1953
Atlanta	*P. vivax*	204	Cooper et al. 1953
Atlanta	*P. vivax*	59	Hernandez et al. 1953
Atlanta	**	11	Myatt, Hernandez, and Coatney 1953
Atlanta	**	35	Myatt et al. 1953
Atlanta	*P. vivax*	51	Coatney et al. 1963
Atlanta	*P. vivax*	24	Coatney, Contacos, and Lunn 1964
Atlanta	*P. vivax*	15	Contacos, Elder, and Coatney 1963
Atlanta	*P. falciparum*	14	Contacos, Elder, and Coatney 1963
Atlanta	*P. falciparum*	14	Contacos et al. 1964
Atlanta	*P. falciparum*	21	Lunn et al. 1964
Atlanta	*P. falciparum*	3	Contacos et al. 1965
Atlanta	*P. vivax*	5	Lunn et al. 1965
Atlanta	*P. ovale*	1	Chin and Contacos 1966
Atlanta	*P. ovale*	8	Chin, Contacos, and Buxbaum 1966
Atlanta	*P. vivax*	4	Chin, Contacos, and Coatney 1966
Atlanta	*P. vivax*	15	Lunn et al. 1966
Atlanta[2]	*P. vivax*	10	Tobie et al. 1966
Atlanta	*P. falciparum*	40	Chin et al. 1966
Atlanta	*P. falciparum*	5	Lunn et al. 1966
Atlanta	*P. falciparum*	9	Chin et al. 1967
Atlanta	*P. vivax*	4	Chin, Coatney, and King 1967
Atlanta	*P. vivax*	6	Chin et al. 1967
	P. falciparum	1	Contacos and Collins 1968
Atlanta	*P. falciparum*	2	Collins et al. 1968[1]
Atlanta	*P. ovale*	5	Chin and Coatney 1971
Atlanta	*P. vivax*	18	Collins et al. 1972[2]
Atlanta	*P. vivax*	10	Contacos et al. 1972
Atlanta	*P. falciparum*	25	Collins et al. 1973c[2]
Atlanta	*P. vivax*	57	Miller et al. 1974
	Subtotal	702	
Jessup[3]	*P. falciparum*	33	Clyde et al. 1971
	Subtotal	33	
Stateville[4]	*P. falciparum*	133	Dern et al. 1955
Stateville	**	81	Dern et al. 1955

Table 7.1 (Continued)

Location		Number of Prisoners	Reference
Stateville	*P. falciparum*	23	Powell, Brewer, and Alving 1963
Stateville	*P. falciparum*	23	DeGowin and Powell 1965
Stateville	*P. falciparum*	22	Powell, DeGowin, and Eppes 1965
Stateville	*P. falciparum*	16	Powell and Brewer 1965
Stateville	*P. falciparum*	50	Powell and Brewer 1967
Stateville	*P. falciparum*	29	Rieckmann et al. 1971
Stateville	*P. falciparum*	35	Willerson et al. 1972a
Stateville	*P. falciparum*	42	Willerson et al. 1972b
	Subtotal	**508**	
Total		**1233**	

**testing antimalarial drug effects without plasmodial infection.
[1]Atlanta State Penitentiary.
[2]Tobie and associates (1966) indicate that some of the 10 prisoners (unspecified) came from Lewisberg, Pennsylvania.
[3]Maryland House of Correction, Jessup.
[4]Illinois State Penitentiary, Statesville Branch.

human, from an infected human to an uninfected human, and from an infected human to an uninfected monkey. Parasite transfer was also accomplished through direct blood inoculation, using the same dyads. Researchers also used the opportunity to test the effectiveness of antimalarials in treating inmates they infected with monkey malarias.

The researchers considered the most important findings to be cases in which it was possible to effectively transfer malaria from a monkey to a human through a mosquito bite. Although conducted under controlled circumstances, these cases were considered to be examples of "natural infections," because they mimicked a potential scenario outside the laboratory. Specifically, the cases of lateral transfer raised the possibility that some of the wild-primate malarias might already exist as zoonotic infections in human populations. Seven types of nonhuman primate malarias were used. Table 7.2 provides a list of the available published experiments that have been identified in the literature. However, information on unsuccessful attempts was not always published.

The bulk of the studies involved *P. cynomolgi* following Eyles's infection. Today, *P. cynomolgi* is known to occur naturally in eight species of monkeys (*Macaca cyclopis, M. fascicularis, M. mulatta, M. nemestrina, M. radiata, M. sinica, Trachypithecus cristatus,* and *Semnopithecus entellus*; see Appendix 1 for a complete reference list). Eyles's accidental infection was surprising, because before 1960, several unsuccessful attempts

Table 7.2 Atlanta-Inmates Experiments with Wild-Primate Malarias

Malarial Form	Number of Inmates	Reference
P. brasilianum	9	Contacos et al. 1963[2]
	10	Contacos and Coatney 1963
	6	Coatney et al. 1971
P. coatneyi	1	Contacos and Coatney 1963
	8	Coatney et al. 1971[5]
P. cynomolgi	2	Eyles, Coatney, and Getz 1960
	25	Beye et al. 1961
	9	Schmidt, Greenland, and Genther 1961[1]
	18	Coatney et al. 1961
	19	Contacos et al. 1962
	3	Contacos and Coatney 1963
	5	Tobie and Coatney 1964[3]
	1	Bennett and Warren 1965[1]
P. cynomolgi	5	Kuvin et al. 1962a; Kuvin et al. 1962b; Tobie et al. 1962
P. vivax + *P. cynomolgi*	4	
P. fieldii	1	Contacos and Coatney 1963
P. fragile	2	Coatney et al. 1971
P. gonderi	3	Contacos and Coatney 1963
P. hylobati	2	Coatney et al. 1971
P. inui	4	Contacos and Coatney 1963
	7	Coatney et al. 1966; Coatney et al. 1971
P. jefferyi	2	Coatney et al. 1971
P. knowlesi	7	Chin et al. 1965
	8	Chin et al. 1968
	1	Coatney et al. 1971
P. schwetzi	13	Coatney et al. 1971; Contacos et al. 1970
P. simium	Unkn	Contacos 1970
	161[4]	

[1]The study does not specify the identity of the study group. However, it can be reasonably concluded that it is Atlanta Penitentiary inmates, since this is the population cited by the authors in similar studies, and it is the only known place where experimental simian infections of human beings took place. Although the study does not specify that the study group was the Atlanta Penitentiary inmates, one can reasonably conclude that this was the population, given that the authors cite the Atlanta prison in similar studies.

[2]See Coatney 1971.

[3]The total number of patients in this study was 7. However, 2 patients included in this study were also part of the study conducted by Coatney and associates 1961.

[4]This number is based on published studies but is an underestimate. By September 1967, Coatney (1968) stated that 204 inmates had been involved in testing simian malaria. By that number, there would have been at least 235 inmates involved. Contacos (1970) indicates that the number of infections established were over 100, with *P. cynomolgi*, 25 with *P. brasilianum*, 20 with *P. knowlesi*, 11 with *P. schwetzi*, and 7 with *P. inui*. Note that Contacos figures indicate successful infections, not the number of attempts made to infect prisoners.

[5]Coatney and associates (1971) indicate that attempts were made on 9 prisoners in 1963 and 1971.

had been made to transfer *P. cynomolgi* to humans. Sinton and colleagues (1938) were unsuccessful in an attempt to transmit *P. cynomolgi* to a human using parasitized blood. Another unsuccessful study involved attempting to infect patients with neurosyphilis with *P. cynomolgi, P. knowlesi,* and *P. inui* by mosquito bite (Coggeshall 1941). During the general time period of Sinton's second study, malaria was being used to treat patients admitted to psychiatric hospitals who had developed advanced-stage syphilis, to be further discussed shortly (for example, Kusch, Milam, and Stratman-Thomas 1936; Meagher 1929[1]). Before 1960 malaria researchers in primate laboratories took few precautions to protect themselves from mosquito bites, because early studies suggested that wild-primate *P. cynomolgi* could not be transmitted to humans.

In 1960 Eyles published an article in *Science* describing his infection, the infection of other workers, and the first monkey malaria experiments on Atlanta inmates (Eyles, Coatney, and Getz 1960). Blood from one of the workers parasitized with *P. cynomolgi* was inoculated into two prisoners, both of whom developed malaria. According to Coatney and colleagues (1971), infections were also reported a primate research facility in Bethesda, Maryland that same year. In addition, two accidental infections occurred in a laboratory in London, and experimental infection was subsequently induced in 12 "volunteers" (Garnham, Molinari, and Shute 1962).

A more extensive study was done shortly afterward involving 25 inmates (Beye et al. 1961). Here the focus was to determine whether *P. cynomolgi* could be transmitted through a mosquito vector. Twenty of the prisoners were exposed to mosquitoes (*Anopheles quadrimaculatus* or *A. freeborni*) carrying *P. cynomolgi bastienelli* originating in Rhesus monkeys *(Macaca mulatta)*. Twelve of the inmates developed infections. An attempt was also made to transfer malaria from one of the infected prisoners into two others by mosquito vector, and one of the two developed malaria. Three additional prisoners received parasitized blood from one of the infected prisoners, but it is unclear from the study whether they developed an infection. Attempts were also made to transfer *P. cynomolgi* malaria from humans into Rhesus monkeys at various stages of the course of the disease. Several of the monkeys did develop infections.

At almost the same time that Eyles was infected with *P. cynomolgi* in Memphis, a primate research worker (RG) at the Institute of Medical Research in Cincinnati also accidentally contracted a malaria that was later determined to also be *P. cynomolgi* (Schmidt, Greenland, and Genther 1961). A study was conducted involving nine individuals who compared infectivity of the B and M strains of *P. cynomolgi*; however, the publication does not state that the individuals involved were Atlanta inmates. They are referred to as "volunteers," which is the term applied to inmates in most of the studies. However, since two of the three authors used "volunteers" from Atlanta prisoners in subsequent studies and the research was funded by NIAID, it can be reasonably assumed that this study involved those inmates. In brief, the study attempted to transmit the malarial strain (*P. c. bastienelli*) accidentally contracted by an individual identified as RG into two volunteers by *A. freeborni* mosquito. The study also attempted to induce *P. cynomolgi* M strain in seven volunteers, also via *A. freeborni*. The B strain was induced in both attempts, but none of the attempts with the M strain were successful. Publications are also available on several additional studies with *P. cynomolgi* in Table 7.2 (Bennett and Warren 1965; Contacos and Coatney 1963; Contacos et al. 1962; Kuvin 1963; Kuvin et al. 1962a, 1962b; Tobie and Coatney 1964; Tobie et al. 1962).

Experiments with *P. knowlesi* were also initiated after a case of a naturally acquired infection. The first study was conducted after an army member who had been in Bangkok was diagnosed with *P. knowlesi* (Chin et al. 1965). Based on morphology, it was initially thought to be *P. malariae*, but Coatney recognized it as *P. knowlesi* (Contacos 1970; Desowitz 1991). The army personnel member's blood was inoculated into seven Atlanta inmates, and all of them became infected. A later study involved mosquito transfer and eight inmates and (Chin et al. 1968), all of whom developed malaria. Two were infected from a monkey to mosquito to human route. One received an inoculation of parasitized blood from one of the first two inmates who had been infected. The remaining five were infected from human to mosquito to human. It is worth mentioning that mistaking *P. knowlesi* for *P. malariae* is the same initial mistake that was made in the recent Bornean outbreak.

Today, the *P. knowlesi* studies of the 1960s and 1970s are prescient. However, the ability of humans to acquire *P. knowlesi* was previously known. In the 1930s *P. knowlesi* was induced in humans as a treatment for neurosyphilis; it initially created a benign form of malaria but was abandoned in favor of *P. vivax* after it proved to become increasingly virulent when cycled through humans through blood-to-blood transfusions (McCutchen 2008). Before the recent outbreaks in Southeast Asia, only a few cases have been documented of infection of humans with *P. knowlesi* under natural conditions (Chin et al. 1965; Jongwutiwes 2004). However, at this point, it is difficult to determine whether or not the rarity of data on *P. knowlesi* in humans represents an actually rarity of occurrence. The recent work by Singh and colleagues (2004) and Cox-Singh and associates (2008) on the misdiagnosis of *P. knowlesi* as human malaria may ultimately prove that *P. knowlesi* is far more common in humans than previously known. If *P. knowlesi* has only recently acquired the ability to effectively infect human beings, it could pose a serious new threat as an emerging infectious disease, particularly given its virulence in humans.

Several other wild-primate malarias were tested on inmates to a lesser extent. Contacos and Coatney (1963) conducted experiments on *P. coatneyi* (n = 1), *P. fieldi* (n = 1), *P. gonderi* (n = 3), and *P. inui* (n = 4). In a later study, Coatney and associates 1966 infected 7 prisoners with *P. inui* though inoculations with parasitized blood. In addition, one study was identified that involved the transmission of African great ape *P. schwetzi* (Contacos et al. 1970). Infection was attempted in 13 prisoners. Three were via a mosquito vector and 10 were through direct inoculation with parasitized chimpanzee blood. For the mosquito vector transmission, 2 of the 3 inmates developed malaria, and 9 of the 10 receiving chimpanzee blood developed malaria. Coatney and associates (1971) also mention several additional studies that were either not published or have been difficult to locate in the literature involving *P. coatneyi, P. fragile, P. hylobati, P. inui*, and *P. jefferyi.*

In addition, attempts were made to infect prisoners with both South American *P. brasilianum* and *P. simium.* Although it is recognized today that both are likely anthroponotic infections of monkeys, based on the knowledge at that time, the researchers believed they were infecting humans with novel wild-primate malaria. Before the Atlanta Penitentiary studies of the 1960s and 1970s, an attempt was made to infect humans with *P. brasilianum* in 1930 (Clark and Dunn 1931). Infection was attempted in 5 individuals, first by means of inoculation of parasitized blood from the black-handed spider monkey (*Ateles geoffroyi*) and then through a mosquito vector. In the other 3, infection was first attempted via mosquito vector, then via the parasitized blood of the brown-headed spider monkey (*Ateles fusciceps*). The mosquito vectors employed were *Anopheles albimanus, A. tarsimaculatus, A. argyritarsis,* and *A. pseudopunctipennis.* None of the experiments were successful. At the time of the study, although it was known that *P. brasilianum* and *P. malariae* were similar, Clark and Dunn (1931) concluded from the negative results of their experiments that the two were not identical and that monkeys did not serve as a reservoir for human malaria. Contacos and colleagues (1963) attempted to repeat the experiment with 9 Atlanta inmates. They also used black-handed spider monkeys (*Ateles geoffroyi*) but used *A. freeborni* as the vector. Five of the 9 patients developed infections. In a later review article of the malaria studies, Contacos and Coatney (1963) mention that attempts were made to infect 10 additional inmates with *P. brasilianum*, but it is not clear whether they were successful.

The first documented case of a human naturally acquiring *P. simium* from a monkey was recorded in Brazil (Deane, Deane, and Neto 1966). A researcher who had been working on an arboreal platform to capture *Anopheles* mosquitoes was infected with *P. simium.* In a review article (Contacos 1970), it was reported that after the Brazilian case, several attempts were made to infect humans with *P. simium* under controlled conditions, but, surprisingly, none were successful.

The use of prisoners in malaria studies has to be understood within the context of the times. In *Malaria Capers* (1991), Desowitz asserts that the prisoners involved in the wild-primate experiments at the Atlanta Penitentiary were all willing volunteers. He describes their motivations as altruistic, evidenced by the fact that they were not compensated for their participation. However, I am skeptical about the voluntary status of the prisoners involved in these studies. It seems naive to assume that inmates (or anyone) would simply volunteer to be subjected to a disease with unknown consequences. Perhaps I am just more cynical than most people, but I do not think that we can assume to know what was in the minds of the prisoners or the conditions that induced them to "volunteer." These studies were conducted before the Tuskegee African American syphilis studies were exposed in 1972, which forever changed regulations for human experimentation. It is worth noting that no malaria prison studies were conducted after 1972, which suggests that the methods would not be supported by ethical standards established in scientific research post-Tuskegee. It would be easy to simply castigate the researchers as behaving unethically, but I think that such a criticism is unfair. These studies were conducted in the context of a broader cultural mindset that made such experimentation seem to be reasonable for the greater good. In addition, members of Eyles's research team, as well as other researchers, put themselves at risk, experimenting on themselves, in order to better understand if wild-primate infections could be established in humans.

Therapeutic Malaria and Neurosyphilis

The history of human experimentation with malaria also includes studies related to the treatment of advanced syphilis. Today, most syphilis cases can be treated with a dose of penicillin, or other antibiotic in the case of penicillin allergy. But before penicillin came into widespread use after World War II, syphilis was a grave health problem. In its late stages, it can manifest as a multiorgan systemic disease affecting the brain, heart, liver, nervous system, blood vessels, eyes, bones, and joints, which can lead to paralysis, blindness, dementia, and death (CDC 2008). The late stage neurological effects lead to what is called "neurosyphilis," whereby sufferers display a suite of psychotic symptoms severe enough to require psychiatric institutionalization (NINDS 2009). As late as the mid-20th century, 20% of patients in mental institutions and 10% of first admissions were patients suffering from neurosyphilis (Frankenburg and Balessarini 2008).

In 1972 a shift in Western consciousness occurred regarding human experimentation when the Tuskegee syphilis study was exposed (Arens 1972). Although penicillin was readily available and a known cure for syphilis since World War II, African American men who were subjects in the Tuskegee study were intentionally deceived about the availability of penicillin and intentionally untreated in order to study the long-term effects of untreated syphilis (Jones 1993). The Tuskegee syphilis studies intersect with a rather murky area of research involving malaria therapy for neurosyphilis. Before penicillin, one of the standard treatments for neurosyphilis was malaria therapy. It was believed that the high fevers of malaria could potentially have an adverse effect on the spirochetes, slowing the progression of the disease (Chernin 1984; Frankenburg and Balessarini 2008).

The use of malaria to treat neurosyphilis was an established procedure from the 1920s until just after World War II, when penicillin became widely available. Although Julius Wagner-Jauregg was awarded the Nobel Prize in Medicine in 1927 for malaria fever therapy for neurosyphilis, many have questioned if malaria therapy ever had any therapeutic value. In Frankenburg and Balessarini's (2008) review, they note that given the health risks of inducing malaria infection, the procedure tended to be used on those less seriously ill, potentially biasing the measures of efficacy. In addition, they make the point that there never were any case control studies conducted on malaria therapy; that is to say, no study was ever conducted that determined if those who underwent malaria therapy fared better than those who did not.

Regardless, with our 21st-century hindsight, we can and should critically evaluate whether malaria therapy was ever efficacious. If, as the evidence suggests, its benefits were questionable, it is equally important that we are able to put ourselves in the shoes of health care professionals at that time. If malaria therapy was prescribed, in good faith, by medical practitioners who believed this treatment was benefiting patients, we should evaluate the research in that context. However, the ethics of using malaria therapy to treat syphilis post-World War II are questionable. Whereas the Tuskegee experiments present a clear-cut case of deceiving and denying patients medical care, the post-World War II malaria therapy studies for neurosyphilis tread a thinner and more ambiguous line. For example, studies published in 1946, 1947, and 1949 involving infection, respectively, of 186, 123, and 54 neurosyphilitic patients from six mental institution fall into a gray area (Young, Ellis, and Stubbs 1946, 1947; Young and Eyles 1949).

After 1950 penicillin came into widespread use and became the standard treatment for syphilis at all stages. However, malaria therapy continued to be used to treat neurosyphilis among institutionalized mentally ill patients. In a number of studies published after 1950, antimalarial drugs were tested in patients with neurosyphilis. The context described suggests that the patients were prescribed therapeutic malaria treatment and that in the course of the treatment, antimalarials were tested. Many of these experiments took place in the mental institution in Milledgeville, Georgia, led by members of the same group of researchers who were testing antimalarials at the nearby Atlanta Penitentiary. Given the availability and known efficacy of penicillin, it does raise the possibility that the primary purpose was to test antimalaria drugs and that malaria therapy was prescribed as a means to facilitate that drug research, rather than to provide the best-known treatment.

At Milledgeville and other mental institutions, research articles document the infection of nearly 500 patients with malaria between 1952 and 1964, *after* penicillin (Collins, Jeffery, and Skinner 1964a, 1964b, 1964c; Jeffery 1952, 1956; Jeffery, Collins, and Skinner 1963; Jeffery and Eyles 1954, 1955; Jeffery, Young and Wilcox 1954; Jeffery et al. 1952; Young 1957). Approximately half the patients were infected with the virulent *P. falciparum,* and half were infected with *P. vivax* (only 15 were infected with *P. malariae).* The choice of plasmodial species also raises suspicions regarding whether the primary motivation was treatment of neurosyphilis or an opportunity for drug testing. By then it had been known for quite some time that many African Americans have resistance to *P. vivax* malaria (owing to high rates of RBC Duffy negativity). Because *P. falciparum* could result in a virulent, life-threatening infection, it was generally avoided in neurosyphilis treatment, and several mental institutions experimented with wild-primate malarias, including great ape *P. reichenowi, P. schwetzi,* and *P. rodhaini,* as well as *P. gonderi* and *P. knowlesi* (Gilks 2001; McCutchan 2008).

The plasmodia used in malaria-therapy were typically *P. ovale* and *P. malariae.* The use of virulent *P. falciparum* is particularly suspect given the history of *P. knowlesi* in malaria therapy. As previously described, early on *P. knowlesi* was used as a purported form of therapeutic malaria but was abandoned, because it became more virulent when cycled through several generations of human hosts (see McCutchan 2008). In a 1960s study at South Carolina State Hospital, nine neurosyphilis patients were infected with a strain of falciparum malaria known to be multidrug-resistant (Jeffery, Collins, and Skinner 1963). It is difficult to imagine a 1960s rationale for treating syphilitic patients with malaria rather than penicillin and, moreover, for treating patients with the most virulent form of malaria when other options were available.

In several of the studies, large numbers of patients were infected with malaria at the same time, which suggests that the goal was experimentation rather than individualized treatment. In one of the studies, neurosyphilitic patients were infected with falciparum malaria and intentionally provided inadequate treatment with antimalarials in order to document the natural duration of the illness (Jeffery and Eyles 1954). In this study 96 patients were infected with falciparum, and it was noted that 23 patients were involved in the study for an average of about 9 months, with 4 of them remaining untreated for more than a year (in Milledgeville, Georgia). In a study on the development of antibodies in malaria, 21 African Americans were infected and underwent malaria therapy. Three were neurosyphilitic patients who were infected with *P. malariae* (Collins, Jeffery, and Skinner 1964a), 10 were neurosyphilitic patients infected with *P. falciparum* (Collins, Jeffery, and Skinner 1964b), and 8 had cerebo-spinal syphilis and were infected with *P. falciparum* (Collins, Jeffery, and Skinner 1964c).

Malaria studies also took place among other vulnerable groups. One disturbing study involved testing antimalarials on children as young as 6 months old. In a 1958–1959 study, over 300 children were given antimalarials to determine how the drugs affected hematocrit levels (Hodgkinson, Courtney, and Haggerty 1961). In this study, 103 children were experimented on who resided at an unidentified institution in the United States, presumably an orphanage. The other 235 children were described as being from East Africa and the Congo. Eleven were under 2 years of age, with the youngest only 6 months old. Although this study did not involve infection with malaria, it is clear at least in the case of the American children that the only purpose of the study was to test the potential adverse effects of malaria. Malaria had been essentially eradicated in the United States by World War II (Humphreys 2001). If any of the African children were afflicted with malaria, it was not stated in the study, and no description was given of therapeutic benefits to the subjects involved. An additional published study in a foreign setting is Young and Moore's (1961) infection of seven "patients" with chloroquine resistant *P. falciparum*, including a 78-pound boy.

One of the ethical dilemmas of the malaria research conducted with the vulnerable populations of prisoners, the mentally ill, orphans, and foreign populations is in whether this information should be used. In 1972, following the exposure of the Tuskegee experiments, Senator Edward Kennedy called for congressional hearings to investigate. This led to the passage of the National Research Act in 1974, which resulted in the establishment of Institutional Review Boards to review proposals for research involving human subjects before they are conducted. Shortly thereafter, the United States Department of Health, Education, and Welfare (now Health and Human Services) created the Belmont Report (Harms 1979), which establishes ethical guidelines to protect human subjects in research. As previously noted, the malaria experiments on prisoners ceased after the exposure of the Tuskegee experiments in 1972.

Although current regulations prevent additional studies of this kind being conducted legally, the question remains whether the results of data obtained through unethical means should be used. One issue to be considered is that, although the malaria prison studies are unethical by today's standards, the researchers were operating legally under the guidelines and restrictions that were in place at that time. There did not appear to be an attempt to hide the methods being used. In fact, in Coatney's 1967 address to the American Society of Tropical Medicine and Hygiene, he praises the U.S. government for legally sanctioning this type of experimentation.

In addition, we should take into consideration that the motivation of the researchers was to address what was perceived as a real health threat to U.S. citizens. It had not been very long since falciparum-, malariae-, and vivax-type malarias had been eradicated on American soil. The potential for the emergence of a new zoonotic form of malaria in human populations was certainly a real threat. The concern was both that wild-primate malaria could escape the laboratory settings and become established in human populations and that zoonotic infections from individuals exposed in foreign settings could be transmitted to the general population.

The concerns in 1960 are not unlike those today with the discovery of *P. knowlesi* in human populations in Southeast Asia. As will be discussed further, infections are being documented not only in Southeast Asia but also in individuals returning from Southeast Asia. Today, the best information that we have about the potential for various wild-primate plasmodia to infect humans comes from the 1960s and early 1970s experiments. This information might be useful in developing predictive models for where future zoonotic infections may occur by mapping the distribution of wild primates near human settlements.

NOTE

1. In the early literature, the terms *paresis* and *general paralysis* are associated with neurosyphilis.

Chapter 8

Future: The Primate Malaria Landscape

Historical ecology provides apt framework for an interdisciplinary study of malaria. Understanding the primate malarias involves understanding interactions of the ecologies, biologies, and behaviors of parasites, vectors, and hosts. The concept of landscape in historical ecology as described by Balee (2006:77) can be understood as a "place of interaction with a temporal dimension that is historical and cultural as it is evolutionary per se, if not more so, upon which past events have inscribed, sometimes subtly on the land." Although historical ecology is in a sense a human ecology, because it involves understanding human behavior and agency in the environment, recently several social theorists have stressed that ecological frameworks should include transspecies actors (for example, Kohn 2007; Latour 2004; Rabinow 1996). For that reason, the landscape of primate malaria should be approached through understanding cross-primate interactions and vector-host relationships in an ecologically interactive sphere.

Agriculture and Malaria

Previous chapters explored the evolutionary history of primate malarias. Although we do not yet know the exact dates of origin for any of the primate malarias, we do know that malaria predates the coming of the primates, with a deep ancestral history of the plasmodia in the common ancestors of birds, reptiles, and amphibians (Cogswell 2000). In approaching malaria from a historical ecological approach, we are concerned not so much with evolutionary origin but with agency in the environment by primate actors that have shaped the history of the expression of malaria.

The contemporary disease ecology of malaria has been shaped by human cultural behaviors, which have also shaped the disease ecology of wild-primate malarias and the interaction of human and wild-primate plasmodia and mosquito vectors. A "big bang" in the history of malaria appears to be linked to the advent of agriculture. The title of this work, *The 10,000 Year Fever*, points to 10,000 years ago as a ballpark figure for a shift in human history from a reliance on hunting, gathering, and fishing as a major means of subsistence to increasing reliance on domestication of plants and animals.

Although "10,000 years ago" is often cited as a benchmark date for the origin of agriculture, its development did not come about spontaneously, suddenly, or uniformly. It is generally accepted that post-Pleistocene ecological changes played a role, with warmer climates and changes in the distributions and prevalence of plant and animal species that were previously relied on by foraging humans. Thus, the adoption of agriculture is not seen so much as a human "invention" that spread from one group to another but as a subsistence strategy that began to be adopted by multiple human groups across the globe. That is not to suggest that early on plants and animals domesticated by one culture were not adopted by another, but rather, for example, the domestication of maize in the New World occurred independently of the domestication of rice in the Old World. The "agricultural revolution" was also not a human universal. Even today, cultures exist that rely on foraging as a primary subsistence mode. It should also be stressed that the adoption of agriculture was a process, preceded by paracultivation and influenced by new technologies. In Africa, iron tools certainly played a role in the intensification of agriculture, as do modern technologies involved in industrialized agriculture.

One of the earliest studies that linked human cultural behaviors to the disease ecology of malaria was Livingstone's 1958 groundbreaking work that described the interrelationship among iron tools, swidden horticulture, malaria vector proliferation, and the emergence of sickle cell trait in tropical Africa (Cormier 2010). He argued that the development of iron tool technology increased the efficiency of clearing forest for agricultural plots. The clearing of forest for agricultural fields created pools of stagnant water exposed to sunlight, which were an ideal breeding ground for vectors of human malaria. Anopheline mosquitoes exploited the new anthropogenic niche and increased in numbers, resulting in much greater exposure of humans to malaria vectors. Livingstone argued that this led to the genetic adaptation of increased frequency of sickle cell trait in tropical Africa, which offers some resistance to falciparum malaria.

As discussed in Chapter 3, 40 years ago and without the benefit of molecular genetics, Livingstone (1971) proposed that human falciparum most likely derived from a lateral transfer from chimpanzees. His rationale was based entirely on ecological inference. Given that agriculture disrupted the natural vegetation, species relying on plants would be driven out locally. Vector-borne parasites would lack hosts and would seek to transfer to new, suitable hosts. He reasoned that chimpanzees were the most likely candidate for the source of human falciparum. In contemporary genetics, multiple studies are implicating African great apes as the possible origin of human falciparum including bonobos (Krief et al. 2010), chimpanzees (Rich et al. 2009) and gorillas (Duval et al. 2010; Liu et al. 2010; Prugnolle et al. 2010). If so, Livingstone's early hypothesis might prove to be correct that anthropogenic environmental changes due to the adoption of agriculture created conditions facilitating a host-switch leading to a novel form of falciparum-type malaria in humans—if not deriving from chimpanzees, then at least from one of the African apes. An additional source of information linking the proliferation of malaria to agriculture is evidence of the relatively recent speciation of the primary *P. falciparum* vector, *Anopheles gambiae* that entomologists have attributed to the changing landscape from a forest zone to an agricultural zone (Coluzzi et al. 1979; Coluzzi, Petrarca, and di Deco 1985).

Historical ecology posits agency as an understudied factor in understanding landscape. HIV-AIDS provides an analogy. Human HIV has been determined to be due to a lateral transfer from chimpanzees (Gao et al. 1999). Although chimpanzee SIV appears to be an ancient disease, it became an epidemic human disease only beginning in the 1980s. Evolutionary variables explain how a host-transfer occurred but not why and when HIV emerged as a human epidemic. The necessary evolutionary conditions were likely just as present in 1980 as they were in 1880, or a thousand years earlier, for a host-switch. But lateral transfers are not only the result of biological susceptibility. The human HIV epidemic, as is well known, is due to multiple interacting factors, including the industrialization of Africa, the rise of the sex worker industry, changing sexual mores, IV drug abuse, the facility of international travel, and blood transfusion technology. Although we cannot know for sure, there is no logical reason to assume that chimpanzee or other wild-primate SIV host-switches occurred only in the late 20th century. Such host-switches may very well have occurred repeatedly in human history, but the sufficient conditions to create a human epidemic may have only recently been in place.

In North and South America, the malarial landscape that was gradually created in the Old World seems to have been replicated and accelerated in the New World with the rapid transformation of the environment from wild forest into settlements and agricultural zones. However, we should back up a step, for such "wild forest" in the New World may also be an artifact of the European entry into the New World. In Livingstone's 1958 article, he proposed a solution to combat African malaria that he called "rejunglization." He cited a source that noted that in "rejungalized" zones in Central Africa, malarial vectors were greatly diminished:

> We felt that it would be interesting to know what would be the condition of things when the rubber trees had grown and the unplanted ravines and swamps had become "rejungalized." We surveyed Mt. Barclay Plantation where the stream borders have grown up with the brush or long grass. After a long search in the streams, we found only two or three larvae, A. mauritianium and A. obscurus. In a pool near a village A. costalis was plentiful (Barber Rice, and Brown 1933:629 in Livingstone 1958). [also note, according to Livingstone, A. costalis = A. gambiae]

Livingstone unequivocally placed the proliferation of malaria with the advent of agriculture. If agricultural zones were to be "rejungalized," he proposed that it would reverse the ecological conditions that allowed *Anopheles* mosquito vector proliferation.

Such reforestation apparently occurred in the New World. Recent work in historical ecology in North and South America has challenged what Denevan has referred to as the "pristine myth" (Balée and Erickson 2006; Denevan 1992a; Heckenberger et al. 2003). The pristine myth argues that 18th-century descriptions of New World as characterized by a preponderance of wild forest did not represent ancient forest conditions but a landscape that had been radically changed owing to infectious disease. Circum-16th-century explorers and later colonists introduced diseases that led to a demographic collapse in the New World, killing off, by conservative estimates, upward of 90% of the indigenous population. The human demographic collapse was also a major ecological event. The sudden, massive depopulation of indigenous Americans also created shifts in the flora and fauna owing to the absence of human environmental agents. In effect, the massive death toll of indigenous Americans, beginning with the introduction of diseases in 1492, created a negative space in the zones previously settled by indigenous people, allowing for forest regrowth. Thus, the wild New World forests documented in the 18th century did not represent a long-standing primitive condition but, rather, conditions created within the last few hundred years—whereby previous human-developed spaces had in effect, "gone wild."

The ecological rebound was short-lived, because as large numbers of colonists arrived and populated the Americas, they also created conditions for the proliferation of malaria. After the initial forest regrowth, the alteration of the landscape by the colonists mimicked the malarial *environment* created with the beginning of agriculture in Africa. A number or researchers (for example, Cleary 2001; Marques 1987) have noted that environmental changes during the colonial period created ideal breeding grounds for vectors. In Humphreys' (2001) history of malaria in the United States, she describes rice, cotton, and sugar cane cultivation as linked to the rise of malaria from colonial times to the early 1940s.

In contemporary times, multiple studies from around the world demonstrate that agricultural regions are associated with the increased rates of malaria. In Africa, examples include studies from Ghana (Klinkenberg et al. 2008), Mali (Diuk-Wasser et al. 2007), Gambia (Lamin et al. 2009), and Nigeria (Oladepo et al. 2010). In Southern Madagascar, malarial transmissions were found to be 150 times higher around irrigated rice fields than in undisturbed habitats (Marrama et al. 2004). Asian examples include Tajikistan (Matthys et al. 2008), Pakistan (Herrel et al. 2004), and Korea (Rueda et al. 2010). In the Amazon region, agricultural practices in combination with logging, hydroelectric dams, mining, and road construction have all been implicated in increases in malaria transmission (Smith 1981; Tadei et al. 1998; see also Yewhalaw et al. 2009). In the Northeast Amazon, *An. marajoara* has recently changed from a minor vector to the major vector of malaria owing to development and alteration of the mosquito breeding grounds (Conn et al. 2002). One of the ironies is that although the adoption of irrigation agriculture can help alleviate hunger in lesser developed areas, the cost in many areas has been an increased malaria burden (Ijumba and Lindsay 2001; Service 1989).

Humans and Wild-Primate Interactions

Landscape transformation through agriculture creates zones where humans and wild primates are more likely to come into contact. For one, deforestation for agriculture and development destroys the natural habitats of wild primates and decreases the distance between humans and wild primates. Under such conditions, wild primates may enter into human areas. In Central and South America, capuchins (*Cebus* spp.) are notorious crop raiders (Baldwin and Baldwin 1976; Balée 1994; Green 1976; Hernández-Comancho and Cooper 1976; Klein and Klein 1976; Moynihan 1976) as are macaques in Southeast Asia and India (Southwick and Siddiqi 1977; Wheatley 1999; Wheatley et al. 2002). Wheatley (1999) has argued that the "natural" habitat of some Old World monkey species may involve close association with humans, including Rhesus monkeys in north-central India, where 86%–88% of the population live in commensal or semicommensal habitats with humans.

In many cultures, wild primates are kept as pets or hunted as food. In South America, there is evidence to suggest that pet-keeping has a direct effect on the prevalence of *P. malariae* in South America. Many indigenous Amazonian peoples not only hunt monkeys for food but also incorporate them into their households as pets (see Cormier 2006 for a review). Such close interactions set up an environment where diseases can be shared. Few studies are available that consider ethnographic information in understanding the distribution of malaria in South America. As will be discussed further, the problem is complicated owing to the way that malaria surveillance is conducted and also to the way in which statistics are regionally compiled (see also Cormier 2010). Those cultures that have been the least Westernized are the most likely to fail to be surveyed adequately. However, these are the very people who live in the closest proximity to wild primates. In addition, culturally specific conditions that could provide insight into malaria transmission are lost when statistics are aggregated. The WHO (2005) estimates that *P. malariae* represents less than 1% of malaria cases in Amazonia; however, the rate among indigenous Amazonians may be much higher.

Few published studies are available on the prevalence of malaria among indigenous Amazonian groups who may be outside the areas of regular malaria surveillance. Those that are available suggest that significant differences may exist between urban and developed rural areas and indigenous people living in tropical forest habitats where data is more difficult to gather. In a study of several indigenous groups in French Guiana, it was found that 38.8% of the Wayana and the Emerillon peoples of the Maroni River basin and 45.4% of the Wayampi and Emerillon people of the Oyapock River basin were seropositive for antisporozoite antibodies against *P. malariae/P. brasilianum* (Volney et al. 2002). The study also tested the sera of 113 local monkeys and found that 73% tested positive for the *P. malariae/P. brasilianum* circumsporozoite peptide. A similar study was conducted in the Xingu Basin of northern Brazil and found that almost 90% of adult Asurini[1] Indians and 100% of adult Metuktire Indians tested had antisporozoite antibodies against *P. malariae/P. brasilianum* (De Arruda et al. 1989). The study authors suspected that keeping monkeys as pets was a likely explanation for the high rates of *P. malariae*. These findings seem at odds with the WHO statistic that less than 1% of malaria in South America is *P. malariae*.

A sylvatic enzootic cycle involving humans and wild primates has been identified for both dengue and yellow fever (Gubler 2002; WHO 2009). A similar process may be occurring with *P. brasilianum* and *P. malariae*. As previously discussed, many malariologists do not believe that the two are separate species; rather, it is designated *P. brasilianum* if found in a monkey and *P. malariae* if found in a human. If such an enzootic cycle is occurring, monkeys serve as a reservoir for human malaria, and likewise humans serve as a reservoir for monkey malaria, with each contributing to the cross-infection of the other.

Rates of 40%–100% of *P. malariae* in some indigenous groups are certainly attention-getting. Various reasons exist for why we do not have better data on malaria in indigenous Amazonians. One is that the WHO compiles aggregate statistics for countries and regions in order to reveal epidemiological patterns in malaria presence, taking a macro rather than a micro approach. This approach is not without merit for it helps to target the areas in most need for intervention. In South America, statistics for the Amazonia include areas contiguous with the rain forest in Bolivia, Brazil, Colombia, Ecuador, French Guiana, Guyana, Peru, Suriname, and Venezuela (WHO 2005). In this region, the WHO 2005 report indicated that 25% of malaria cases were due to *P. falciparum*, and 75% were due to *P. vivax*. In essence, *P. malariae* is invisible in these statistics.

One of the problems in accurately measuring the prevalence of malaria (and relative frequencies of type of plasmodial infection) is that malaria is usually diagnosed through passive case detection (PCD). Active case detection (ACD) is less common and aggressive active case detection (AACD) is much rarer. In passive case detection, an individual is tested for malaria only after reporting symptoms to a health care worker. In 2000 nearly 5 million malaria smears were collected, but 75% were through passive case detection (McCauley 2005). Active case detection involves periodically visiting communities and collecting blood samples from individuals who have had a fever since the last collection round. Aggressive active case detection (AACD) involves taking monthly blood surveys of everyone, whether or not an individual is exhibiting a fever. In Brazil, where most of the rain forest lies, aggressive active case detection has been implemented in only very limited areas, such as the Yanomami program, which has been successful in reducing malaria rates (McCauley 2005). Despite the success of AACD in selected groups, nearly all malaria cases in Brazil are still identified through either active or passive case detection (Da Silva-Nunes and Ferreira 2007).

A major problem with compiling statistics from either active or passive case detection is that many communities are not in regular contact with Western health care services. The Pan American Health Organization (2006) estimates that 80% of indigenous peoples rely on traditional healers and that 40% lack access to biomedical health care. If 75% of the documented cases of malaria are documented only through self-report to a biomedical health care worker, many cases are likely to be missed owing to lack of access and through preference of going to a traditional healer. The PAHO also reports that when biomedical health care is available, cultural barriers such as language differences or mistrust may prevent indigenous people from utilizing these services. And again, people living in the most remote areas are least likely to have contact with biomedical surveillance and are the most likely to have contact with monkey populations. From my own research with the Guajá, a physician sponsored by the CVRD railroad came by once a year. While nurses were available at the post, passive case detection was used. Although the Guajá had a village, at any given time, half of them may be on extended treks in the forest. And even if spending the night in the village, during the day, many are often in the forest foraging. So, many cases may go undetected.

Two additional problems exist with detection. *P. malariae* may be harder to detect by microscopy. In a survey in western Rondônia, Brazil using PCR (polymerase chain reaction), *P. malariae* was found to occur at a rate of 10% (Cavasini et al. 2000). The authors note that 2–3 million GTS (Giemsa-stained thick smears) are examined yearly in Brazil, yielding a rate of *P. malariae* at 0.3%. A similar study in Mato Grosso, Brazil tested the same group and found PCR prevalence of *P. malariae* at 11.9%, whereas GTS showed only 1.2% (Scopel et al. 2004). In addition, because *P. malariae* is a relatively benign form of malaria (Mueller, Zimmerman, and Reeder 2007), individuals infected may not exhibit symptoms severe enough to actively seek treatment, unlike *P. falciparum*, which can be incapacitating. In a review article of studies of asymptomatic malaria in Brazil, the rate of positive results from PCR testing in asymptomatic individuals ranged from 20.4% to 49.5% (Coura, Suárez-Mutis, and Ladeia-Andrade 2006). If individuals with *P. malariae* do not report fever, they will be missed by both active and passive case detection. For even with active case detection, typically only those who have reported fever will be tested. If subclinical cases of *P. malariae* are not identified and treated, then the reservoir for the disease will be maintained.

One final comment on the case of indigenous Amazonians is that current WHO (2005) recommendations for intervention measures to control mosquitoes are likely inadequate and inappropriate for tropical forest dwelling people. Larviciding, indoor residual spraying, and insecticide treated mosquito netting can offer little protection for peoples who spend significant amounts of time in the forest. Such interventions also do not address a key potential problem of forest monkeys serving as malaria amplifiers and reservoirs. The same is certainly true in tropical Africa. If the African apes do prove to be reservoirs for *P. falciparum*, larviciding a dwelling will not be sufficient.

The Cultural Context of Malaria

The physician and medical anthropologist Paul Farmer has argued that malaria is best understood as a disease of poverty, because describing malaria as a "tropical disease" implies geography rather than topography (2001:41). In his own work in Haiti, he found that his patients with malaria were almost exclusively those living in poverty. His point is well taken, because describing malaria as a tropical disease is both historically false and elides socioeconomic factors with geographic locale. Today, the threat of malaria is largely confined to the warm tropical areas of the world, but this is not merely an artifact of ecological conditions but of larger social, economic, and political factors.

One disease analogy that is perhaps useful as a contrast is the mortality rate for measles in the contemporary world. Statistics from the World Health Organization state that in 2007, 20 million people contracted measles, primarily in low-income nations; there were nearly 200,000 deaths from measles globally, which translates to more than 500 deaths per day, more than 20 per hour, and more than 3 per minute (WHO 2008a). A safe and effective vaccine for measles was first licensed in 1963 (Bloch et al. 1985), nearly 50 years ago. We have the technology to rid the world of measles through vaccination, but social, political, and economic conditions have prevented such a simple measure from being implemented among the world's citizens. Given its distribution, one could argue that measles is a "tropical disease," because today it is endemic only in the tropical areas of the world. Obviously this would be a false association. Measles is endemic only among peoples experiencing extreme poverty. A similar case could be made for polio. Jonas Salk in 1955 developed an injectable polio vaccine, and Albert Sabin in 1957 developed an oral polio vaccine (Sabin 1985). Yet, more than 50 years later, polio, which is preventable, remains a serious health threat in some low-income nations[2] (2008b).

The World Health Organization describes malaria as "preventable and curable," and yet approximately half the world's population remains at risk for malaria—for the most part, people living in low-income countries (WHO 2009). Malaria not only is most likely to affect people in low-income nations; it also maintains a negative synergy with high economic costs. The WHO (2009) estimates that in heavy-burden areas, malaria accounts for up to 40% of public health expenditures, 30%–50% of hospital admissions, and up to 60% of outpatient health visits. The WHO also estimates that in high-rate areas, the GDP (gross domestic product) is decreased by as much as 1.3% by malaria alone (primarily in Africa). These statistics are staggering, particularly considering the health burden of HIV-AIDS. Although malaria typically receives less attention than HIV-AIDS, it arguably has taken a greater toll on humanity, particularly given that epidemic malaria has afflicted human populations for thousands of years. Research is ongoing to develop a vaccine for malaria and the development of genetically modified mosquitoes that would be unable to transmit malaria. However, most malaria control measures today involve antimalarial drugs and insecticides. The possibility of reintroducing DDT as a malaria control measure, although controversial, is being seriously debated today.

DDT: Monster or Miracle?

Most of the lay public who are familiar with the term *DDT* have some awareness of its designation as a dangerous pesticide that has been banned. DDT was the first modern pesticide and has been used in malaria control and more broadly in controlling many kinds of insects. DDT was initially seen as almost miraculous until Rachel Carson's publication of *Silent Spring* (1962), which argued that DDT had very serious harmful effects on wildlife. *Silent Spring* was widely read by the general public and arguably was one of the key works that led to the modern environmental movement. Today, Carson's research that led to the banning of DDT in the United States and many other nations is coming under scrutiny. *Silent Spring* has even been called a scientific fraud (Edwards 2004). The World Health Organization has reapproved DDT for indoor spraying and is considering outdoor spraying (WHO 2006). The history of DDT and *Silent Spring* will be reviewed below with attention to two key issues: does DDT pose a threat to the environment and wildlife; and is DDT an effective measure against malaria?

The history of DDT starts with the chemist, Justus von Liebig. In ecology, he is perhaps best known for his "law of the minimum," which states that a plant's development is limited by one essential mineral that is in the relatively shortest supply (Paris 1992). In anthropology, the idea of "limiting factors" was adapted to suggest that poverty of a local resource, such as poor soils (for example, Meggers 1957, 1973) or insufficient protein (for instance, Good 1987; Gross 1975; Harris 1974) affect population size, extent of sedentism, cultural behaviors and values, or general degree of cultural development.

Liebig's contribution to DDT was his discovery of chloral in 1832 through the chlorination of ethanol (Fürstenwerth 2003). He combined chloral with water to form chloral hydrate, which became one of the first sedatives (Snelders, Kaplan, and Pieters, 2006). Chloral hydrate is also the original drug used to "slip a Mickey Finn" or "slip a mickey," meaning to spike a person's drink with a drug. Mickey Finn was the owner of the Lone Star Saloon in Chicago at the turn of the 20th century who is alleged to have incapacitated customers by slipping choral hydrate into their drinks and then robbing them (Algren 1951; Bechtel and Holstege 2007).

DDT (Dichloro-diphenyl-trichlorethane) was first synthesized in 1874 in Germany by Othmar Zeidler as part of his doctoral work in chemistry by combining chlorobenzene and chloral in the presence of sulfuric acid (Fürstenwerth 2003; Metcalf 1973). Over 60 years passed before its insecticidal properties were discovered in 1939 by Paul Müller, an employee of the Swiss company Geigy AG (Fürstenwerth 2003; Metcalf 1973). DDT came into widespread use during World War II, when it was applied directly to the body as a dust and in a solution to clothing in order to combat vector-borne diseases such as plague, typhus, malaria, and yellow fever (Jukes 1963; Metcalf 1973). Müller was subsequently awarded the Nobel Prize in Medicine in 1948 (Fürstenwerth 2003; Metcalf 1973).

Beginning in 1955, the World Health Assembly (of the WHO) began a global malaria eradication program relying on the use of DDT (Matteson 1999). It was credited with saving millions of lives and eradicating or drastically reducing malaria in 37 countries (Matteson 1999; Metcalf 1973). However, according to Matteson, by 1969 the program was viewed as unsustainable by the WHA owing not only to a lack of resources but also to mosquito resistance to DDT and increasing problems of drug resistance (Matteson 1999). In fact, the WHA in 1955 understood beforehand that insecticide resistance would occur and recommended that DDT spraying be done rapidly to attempt to eradicate the malaria before the mosquitoes had a chance to develop resistance (Litsios 2005). In 1978, the WHA shifted responsibility to local health care services and redefined the goal to be "control" rather than "eradication" of malaria (Matteson 1999). However, between the time of the initiation of the WHA malaria eradication program and its decentralization, public concern was gathering over the potential hazards of DDT, owing to the publication of Rachel Carson's *Silent Spring* in 1962.

Researchers in multiple fields have cited Rachel Carson's publication of *Silent Spring* as the genesis of the modern environmental movement (for example, Bosselman and Tarlock, 1993; Carfaro 2001; Edwards 2004; Kroeber 1997). Kroeber (1997:317) describes the impact:

> Most of the popular support for her position was generated by the book itself. She did not merely give focus to well-defined, widely recognized beliefs: she provided a point around which new mental attitudes could begin to crystallize politically. . . . Carson accused American civilization, the essence of modern civilization, of having declared war on the natural world.

Environmentalism certainly existed in many forms prior to *Silent Spring* including Thoreauian followers advocating self-discovery in simple, natural surroundings and organizations concerned with preservation and conservation such as the Sierra Club and the Audubon Society. Carson's work was a watershed event because it represented a shift in the public consciousness. The idea of conservation had previously involved more of an appreciation of nature. With Carson's work, the idea conservation came to mean addressing imminent threats to nature and to human existence.

One of the key issues raised by Carson was her claim that Congress had been keeping records on the effects of DDT in birds that had not been made public. She went on to describe various problems with DDT exposure in birds including thinning of egg shells and declines in various bird populations. Most notably, she linked DDT to the decline of the bald eagle, which proved to particularly alarming to the public, perhaps because of its status as a national symbol. But perhaps the most influential aspect of her book was that it suggested a betrayal on the part of industry of the public trust.

As Kroeber (1997) also notes, although criticism of the role of industries in environmental pollution is now commonplace, it was not in 1962. Although the public embraced Carson's book as a call to action, she met with attacks from chemical companies, government agencies, and some scientific quarters. Even before *Silent Spring* was published, there was a well-financed and well-orchestrated attempt to suppress its publication by the chemical industry. She was even accused of supporting a communist plot to undermine third world countries. To put that accusation in historical context, her book was published not long after the days of McCarthyism and the beginning of the cold war era. Just after World War II, a number of events created extreme concern about the spread of communism across the globe, including Soviet testing of an atomic bomb, the rise of Mao Zedong's communist army, U.S. involvement in the Korean War. The year before *Silent Spring* was published, construction began on the Berlin Wall, the Bay of Pigs invasion failed, and John F. Kennedy began a major build-up of U.S. troops in Viet Nam. So, in that climate, accusing Carson of being involved in a communist conspiracy was a potentially ruinous form of character assassination.

Carson was also the subject of numerous sexist derisions of her work involving gender stereotypes of the time that dismissed her work by simultaneously characterized her as being too feminine and too unfeminine. Corbett (2001) has described some of the comments made in the popular press, which disparaged her as being shrill, emotional, or simply being merely a woman:

> "simply a long emotional attack" "emotional and one-sided" "an obscurantist appeal to the emotions" *National Review* (1962)
>
> "emotions are running so high that it is difficult to uncover the facts" *Popular Mechanics* (1963)
>
> Carson's "current outburst" *Today's Health* (1963)

> "high-pitched sequences of anxieties" *Chemical & Engineering News* (1962)
>
> "shy female bachelor biologist" *Saturday Review* (1963)
>
> "the author is a professional journalist—not a scientist in the field" *Atlantic Monthly* (1963)
>
> "a woman who has written a book" *Farm Journal* (1962)
>
> "unfair, one-sided and hysterically over-emphatic" *Time* (1962)
>
> "Many scientists sympathize with Miss Carson's love of wildlife, and even her mystical attachment to the balance of nature. But they fear her emotional and inaccurate outburst in *Silent Spring* may do harm by alarming the nontechnical public, while doing no good for the things she loves" *Time* (1962).

Despite these depictions of Carson as a shy/hysterical, mystic/communist, woman/bachelor, or simply "a woman who has written a book," there was a huge public swelling of support for *Silent Spring*. It was also taken seriously enough by the chemical industries that she was threatened with lawsuits. In the wake of her book, a government panel was formed in 1963 to investigate DDT, and Carson testified. The result of the inquiry directly led to the creation of the U.S. Environmental Protection Agency and to the U.S. ban on DDT.

However, the more serious criticism of *Silent Spring* that is relevant today has come from scientific quarters. Even during her time, some were questioning the research that Carson cited in her book and whether or not she had reported that research accurately. Recently, those questions have again surfaced, and scrutiny has intensified over the claims made over the years regarding the potential deleterious health effects and ecological effects of DDT. The two key questions regarding DDT are (1) does it actual pose health risks to humans, other species, and the environment and(2) is DDT an effective intervention for malaria?

First, DDT is clearly toxic to some species of life. It is, of course, used for that very purpose, to kill mosquitoes and other insects that have potential to threaten the human food supply and to transmit insect-borne diseases. The question is whether or not it is toxic to other forms of life, including humans, and how its ability to kill insects may affect the ecological chain in areas where it is used. DDT is classified as a POP, Persistent Organic Pollutant, along with 11 other agents that are sometimes referred to as the "dirty dozen" (aldrin, chlordane, DDT, dieldrin, endrin, heptachlor, hexachloro-benzens, mirex, toxaphane, polycholorinated biphenyls [PCBs], dixons, and furans) (Adeola 2004). Such POPs are resistant to breakdown by natural processes. According to Adeola

(2004), DDT can remain effective in a local environment for up to 12 years. If you were a farmer and your problem was agricultural pests, DDT was good news. It was relatively inexpensive and easy to apply, with the added benefit that a one-time spraying could protect crops for many years.

The POPs, by their nature, are characterized by two related processes: bioaccumulation and the "grasshopper's effect," which have been described by Adeola (2004). Bioaccumulation means that because DDT does not break down easily, so repeated use can lead to a build-up of the chemical in a local environment. The grasshopper's effect is also related to the persistence of DDT in the environment, but with opposite effects, because it does not stay put. DDT is highly mobile and can be transported far from its intended area of use through runoff into local and remote water systems. Through rain and evaporation, POPs can "hop" and become airbound in warm tropical climates and condense on land in colder zones. Since World War II, DDT has become incorporated into the bodies of virtually every human being on the planet. It has been detected in polar bears, penguins, and even in the breast milk of Inuit women. It should give us all pause to consider that DDT campaigns in the tropical forests of Africa have resulted in distant Inuit women in the Arctic delivering a measurable dose of DDT to their infants when they breastfeed them.

However, in weighing the costs and benefits of DDT use, we must also consider whether or not DDT actually poses a health risk to human life or health. Edwards (2004), who is a proponent of DDT use, describes two early studies that may indicate that the health risks to humans may have been overestimated. In one two-year study, "volunteers" were asked to ingest 30 milligrams of DDT every day. That dosage is approximately 1,000 times the highest level ever recorded in human exposure. No ill health effects were discerned during the study period or several years later. Ethically, this presents a catch-22. For my part, it is appalling to consider that DDT campaigns have placed that chemical in the mouths of nursing infant Arctic children. Yet, we should also be very acutely aware of the massive toll that malaria today takes on human life. WHO (2010) statistics indicate that in 2008, malaria resulted in nearly *one million* human deaths, particularly among African children. The WHO estimates that an African child dies of malaria *every 30–45 seconds*, and the disease accounts for 20% of all childhood deaths in Africa. Should we be more concerned with the immediate potential benefits of DDT to our fellow human citizens or the fuzzier potential long-term consequences?

Whether or not DDT is actually effective in eliminating malaria is debatable. The United States is often cited as an example for how DDT spraying eradicated the disease. However as Humphreys (2001) makes clear in her history of malaria in the United States, malaria had virtually disappeared prior to DDT spraying. Malaria control was achieved through environmental means, draining swamps in the South and the Tennessee Valley Authority's engineering design of dams that discouraged mosquito breeding. Many of the areas where malaria thrives today are tropical rain forests. The environmental engineering that was effective in the United States cannot be applied to the rain forest; we cannot simply drain the water from rain forest habitats. In addition, potential unintended consequences in the ecosystem may occur with DDT use. In an early 1960s DDT campaign in Kuala Lumpur, malaria workers were ejected from the community, because the program caused the roofs of the residents houses to collapse within a month of spraying. According to Desowitz (1991), what had happened was that DDT spraying killed a wasp species that preyed on a caterpillar species that ate palm fronds. Elimination of the wasps allowed the caterpillars to multiply and destroy homes.

Although DDT has clearly demonstrated its ability to control mosquitoes in the short term, they rapidly develop resistance. Mosquitoes develop resistance to DDT in two ways: biological and behavioral. Biological resistance involves population changes in mosquitoes exposed to DDT that either increase their ability to detoxify the chemical or involve the development of insensitivity to the mechanism of action (Ranson et al. 2000). Behavioral changes can also occur so that mosquitoes are selected for that display avoidance of DDT. The problem is not just with DDT, but with all insecticides. The insecticide permethrin is often sprayed or used to treat bednets. In the Ivory Coast, Burkina Faso, and Benin, *A. gambiae* populations have developed resistance to permethrin (Chandre et al. 1999). In Kenya, permethrin resistance appeared in *A. gambiae* only one year after implementing a permethrin bednet program (Vululu et al. 1994; Vululu et al. 1999). Resistance of *Aedes aegypti* to DDT, which is a vector for dengue and yellow fever, developed in the Caribbean as early as 1955, and pyrethroid resistance has been reported in Puerto Rico, the Dominican Republic, the Virgin Islands, Cuba, Martinique, and the Cayman Islands (Harris, Rajatileka, and Ranson 2010).

In the United States and Europe, malaria was eradicated in the 20th century, but these areas did not have endemic wild-primate populations as seen in tropical Asia, South America, and Africa. McCutchan (2008) has commented that in the long term, we have no way of eradicating malaria when wild-primate reservoirs are present. No matter how well we protect humans, the transmission cycle can continue in wild primates. He adds that the issue is not only that of wild primates harboring the plasmodia; we must also consider that the mosquitoes that affect wild primates have different habitats, thereby nullifying the effects of bednets and indoor pesticides. To reiterate in case it is unclear, what the WHO endorsed in 2006 is limited use of DDT spraying inside human dwellings, including the use of insecticide treated mosquito netting. The WHO has judged that indoor DDT spraying does not pose a significant health risk to humans. That may very well we justified. But, McCutchan's point is the WHO is only narrowly addressing the problem. Although indoor spraying of human dwellings with DDT or use of insecticide-treated bednetting may very well have a beneficial effect, such interventions fail to recognize that humans do not live only in their dwellings or their mosquito bednets but also live outside their household dwellings, farming or extracting forest resources that bring them into contact with the larger ecological milieu, that puts wild primates, humans, and potential plasmodial shared vectors into contact.

Malaria as an Emerging Disease

The close genetic relationships among primate species has allowed a number of infectious organisms to jump from their natural hosts to other primate species. Some of the diseases that have been documented in wild primates that are of human origin are scabies (chimpanzees, gorillas), yaws (baboons), schistosomiasis (baboons), *Giardia intestinalis* (howler monkeys), pneumococcal pneumonia (great apes), polio (orangutans, chimpanzees), and tuberculosis and measles, which infect a wide variety of wild primates (Renquist and Whitney 1987; Stuart et al. 1998; Wallis and Lee 1999).

One of the most dangerous host-switching primate diseases are the herpes viruses. Although they tend to be relatively benign in their natural hosts, they can cause serious illness when they switch hosts (Eberle and Hilliard 1995). Rhesus macaques, as previously described, are common laboratory animals for medical research; the CDC estimates that at least 70% of macaques harbor the herpes B virus (*Herpesvirus simiae*) (Centers for Disease Control 1998). When humans contract the virus, it leads to lethal encephalitis in approximately 78% of the people who develop symptoms (CDC 1987a, 1987b; Fiennes 1967). Owl monkeys can contract human herpes virus (*Herpesvirus hominis*), which produces severe disease in them (Weller 1994). Spider monkeys (*Ateles* spp.) are asymptomatically infected with *Herpesvirus ateles*, but it is typically lethal in owl monkeys (*Aotus* spp.) and can cause lymphoma when experimentally induced in tamarins, marmosets, and capuchins (King 1994; Weller 1994). Humans have been documented to be infected with wild-primate diseases including howler monkey tapeworms (Stuart et al. 1998) and spumavirus (Centers for Disease Control 1997).

The most well-known host-switching disease is the origination of human HIV in the common chimpanzee (*Pan troglodytes troglodytes*) (Gao et al. 1999). A nonpathogenic AIDS-like virus has also been found in Brazilian Amazonian Indians that may have been transmitted from Neotropical monkeys to humans (see Feldman 1990, quoting Francis Black). The natural hosts of the ebola filoviruses is unclear, but they are able to infect both humans and wild primates, as are two other viral hemorrhagic fevers, Mapucho and Lassa (Peters et al. 1993).

In 2004, the year that I began researching this book, a large-scale outbreak of malaria in Borneo was discovered to be due to wild primate *P. knowlesi.* It was initially misdiagnosed as *P. malariae* by blood smear, but DNA testing was conducted owing to unusual severity of symptoms, and it was discovered to be *P. knowlesi,* with infection rates as high as 83.7% of malaria cases in one hospital (Cox-Singh et al. 2008; Singh et al. 2004). It is unknown if the 2004 Borneo outbreak was the genesis of human infection with wild-primate *P. knowlesi,* or if 2004 only represented the first accurate diagnosis of a *P. knowlesi* outbreak in humans. It has since been dubbed the "fifth human malaria" (McCutchan 2008; White 2008). I would predict that *P. cynomolgi* may very well prove to be the sixth. It is a natural host of eight different primate species (see Appendix 1) and was first observed to be transmissible from monkeys to humans via a mosquito vector over 50 years ago (Eyles, Coatney, and Getz 1960).

We should also be very concerned with the documentation of *P. falciparum* in the wild in two species of howler monkeys in the New World and (Duarte et al. 2008) and in bonobos, chimpanzees, and gorillas in the Old World (Duval et al. 2010; Krief et al. 2010; Liu et al. 2010; Prugnolle et al. 2010). Multiple Neotropical monkey species have been demonstrated to be susceptible to *P. falciparum* (see Appendix 2), but Duarte and associates (2008) provide the only documented case of *P. falciparum* in wild Neotropical monkeys. Best-case scenario: the study represents only a few isolated cases of anthroponotically acquired *P. falciparum* in monkeys. The implications of the worst-case scenario are chilling. If wild Neotropical monkeys are becoming a reservoir of *P. falciparum,* potentially leading to South America mirroring Africa in *P. falciparum* deaths. And worse, it may be far more difficult to control *P. falciparum* in South America if a patent wild-primate reservoir exists. In Africa, the great apes may have been reservoirs of *P. falciparum* for millennia, which may partially explain why it has remained so difficult to control in human populations.

In conclusion, if 10,000 years ago was one benchmark in the history of malaria, we may be at another benchmark today. One the one hand, there is reason to be optimistic, because with the *RollBack Malaria* program, there is more attention to the global crisis of malaria than there ever has been in the past. However, on the other hand, deforestation and development are radically changing wild-primate habitats, and so wild primates continue to encroach into human-occupied areas to survive. Given the propensity of malaria to host-switch, we have the possibility of additional wild-primate malarias laterally transferring between species and humans. If an effective solution to the malaria problem is to be found, it will have to include not only intervention in human populations but address primates as a whole.

Notes

1. More commonly spelled Assurini.
2. According to the WHO (2008b), four countries remain polio-endemic: Afghanistan, India, Nigeria, and Pakistan.

Appendix 1

Plasmodia Parasites and Their Natural Primate Hosts

Malaria Form	Natural Hosts	References
Falciparum Type		
P. billbrayi sp. n.	*Pan troglodytes*	Krief et al. 2010
P. billcollensi sp. n.	*Pan troglodytes*	Krief et al. 2010
P. falciparum	*Gorilla gorilla* (western)	Duval et al. 2010; Liu et al. 2010; Prugnolle et al. 2010
	Homo sapiens	
	Pan troglodytes	Duval et al. 2010
P. cf. falicparum	*Pan paniscus*	Krief et al. 2010
P. gaboni sp. n.	*Pan troglodytes*	Duval et al. 2010; Kaiser et al. 2010; Ollomo et al. 2009; Prugnolle et al. 2010
P. GorA sp. n.	*Gorilla gorilla* (western)	Prugnolle et al. 2010
P. GorB sp. n.	*Gorilla gorilla* (western)	Prognolle et al. 2010
P. reichenowi	*Gorilla gorilla* s.l.[1]	Bray 1963; Cogswell 2000; Warren and Wharton 1963
	Pan troglodytes[2]	Bray 1963; Coatney et al. 1971; Cogswell 2000; Duval et al. 2010; Kaiser et al. 2010; Krief et al. 2010; Prugnolle et al. 2010; Rich et al. 2009; Warren and Wharton 1963

Vivax Type		
P. coatneyi	*Macaca fascicularis*[3]	Coatney et al. 1971; Cogswell 2000; Escalante, Barrio, and Ayala 2005; Eyles 1963; Eyles et al. 1963; Fooden 1994; Garnham 1963; Seethamchai et al. 2008; Warren and Wharton 1963
	Macaca nemestrina	Putaporntip et al. 2010
P. cynomolgi[4]	*Macaca cyclopis*	Coatney et al. 1971; Cogswell 2000; Contacos and Coatney 1963; Eyles 1963; Warren and Wharton 1963
	Macaca fascicularis	Bennett and Warren 1965; Coatney 1971; Coatney et al. 1971; Cogswell 2000; Contacos and Coatney 1963; Escalante, Barrio, and Ayala 2005; Eyles 1963; Fooden 1994; Garnham 1963; Howard and Cabrera 1961; Warren and Wharton 1963
	Macaca mulatta	Coatney et al. 1971; Cogswell 2000; Escalante, Barrio, and Ayala 2005; Fooden 1994; Schmidt et al. 1961; Warren and Wharton 1963
	Macaca nemestrina	Coatney et al. 1971; Cogswell 2000; Contacos and Coatney 1963; Escalante, Barrio, and Ayala 2005; Eyles 1963; Fooden 1994; Garnham 1963; Warren and Wharton 1963
	Macaca radiata	Coatney et al. 1971; Cogswell 2000; Escalante, Barrio, and Ayala 2005; Eyles 1963; Fooden 1994; Garnham 1963; Warren and Wharton 1963
	Macaca sinica	Coatney et al. 1971; Cogswell 2000; Dissanaike 1965; Escalante, Barrio, and Ayala 2005; Fooden 1994
	Trachypithecus cristatus[5]	Coatney et al. 1971; Escalante, Barrio, and Ayala 2005; Eyles 1963; Garnham 1963; Warren and Wharton 1963
	Semnopithecus entellus[6]	Coatney et al. 1971; Dissanaike 1965; Escalante, Barrio, and Ayala 2005; Fooden 1994
P. fieldi	*Macaca fascicularis*	Cogswell 2000; Escalante, Barrio, and Ayala 2005; Warren and Wharton 1963
	Macaca nemestrina	Coatney et al. 1971; Cogswell 2000; Collins, Warren, and Galland 1999; Contacos and Coatney 1963; Escalante, Barrio, and Ayala 2005; Eyles 1963; Fooden 1994; Garnham 1963; Warren and Wharton 1963

P. *fragile*[7]	*Macaca radiata*	Coatney et al. 1971; Cogswell 2000; Collins et al. 1990b; Escalante, Barrio, and Ayala 2005; Eyles 1963; Puri and Dutta 1994
	Macaca sinica	Coatney et al. 1971; Cogswell 2000; Collins et al. 1990b; Fooden 1994; Puri and Dutta 1994
P. *gonderi*[8]	*Cercocebus atys*	Coatney et al. 1971; Cogswell 2000; Escalante, Barrio, and Ayala 2005; Eyles 1963; Warren and Wharton 1963
	Cercocebus agilis[9]	Coatney et al. 1971; Cogswell 2000; Eyles 1963; Warren and Wharton 1963
	Cercocebus sp. undet.	Contacos and Coatney 1963; Garnham 1963
	Lophocebus aterimus[10]	Coatney et al. 1971; Cogswell 2000; Eyles 1963; Warren and Wharton 1963
	Mandrillus leucophaeus	Coatney et al. 1971; Cogswell 2000; Contacos and Coatney 1963; Garnham 1963; Warren and Wharton 1963
P. *hylobati*	*Hylobates moloch*	Coatney et al. 1971; Cogswell 2000; Collins et al. 1972
	Hylobates sp. undet.	Garnham 1963
P. *inui*[11]	*Macaca assamensis*	Garnham 1963
	Macaca cyclopis	Coatney et al. 1971; Cogswell 2000; Contacos and Coatney 1963; Escalante, Barrio, and Ayala 2005; Fooden 1994; Warren and Wharton 1963
	Macaca fascicularis	Coatney et al. 1971; Cogswell 2000; Contacos and Coatney 1963; Donovan et al. 1983; Escalante, Barrio, and Ayala 2005; Eyles 1963; Eyles and Warren 1962; Fooden 1994; Garnham 1963; Otsuru and Sekikawa 1968; Putaporntip et al. 2010; Schmidt et al. 1980; Seethamchai et al. 2008; Warren and Wharton 1963; Warren et al. 1970
	Macaca mulatta	Coatney 1971; Coatney et al. 1971; Cogswell 2000; Contacos and Coatney 1963; Escalante, Barrio, and Ayala 2005; Fooden 1994; Garnham 1963; Schmidt et al. 1961; Warren and Wharton 1963
	Macaca nemestrina	Coatney et al. 1971; Cogswell 2000; Contacos and Coatney 1963; Escalante, Barrio, and Ayala 2005; Eyles 1963; Eyles and Warren 1962; Fooden 1994; Garnham 1963; Putaporntip et al. 2010; Warren and Wharton 1963

	Macaca nigra[12]	Coatney et al. 1971; Cogswell 2000; Eyles 1963; Eyles and Warren 1962; Fooden 1994; Garnham 1963; Warren and Wharton 1963
	Macaca radiata	Coatney et al. 1971; Cogswell 2000; Contacos and Coatney 1963; Escalante, Barrio, and Ayala 2005; Eyles and Warren 1962; Fooden 1994; Garnham 1963; Warren and Wharton 1963
	Macaca sinica	Dissanaike 1965; Escalante, Barrio, and Ayala 2005; Fooden 1994
	Trachypithecus cristatus	Coatney et al. 1971; Warren and Wharton 1963
	Presbytis melalophos	Warren and Wharton 1963
	Trachypithecus obscurus[13]	Coatney et al. 1971; Warren and Wharton 1963
	Presbytis sp. undet.	Garnham 1963
P. knowlesi	*Macaca cyclopis*	Warren and Wharton 1963
	Macaca fasicularis	Coatney 1971; Coatney and Contacos 1963; Coatney et al. 1971; Cogswell 2000; Escalante, Barrio, and Ayala 2005; Eyles 1963; Fooden 1994; Garnham 1963; Putaporntip et al. 2010; Schmidt et al. 1977b; Seethamchai et al. 2008; Vythilingham et al. 2008; Warren and Wharton 1963; Warren et al. 1970
	Macaca nemestrina	Coatney et al. 1971; Cogswell 2000; Contacos and Coatney 1963; Escalante, Barrio, and Ayala 2005; Eyles 1963; Garnham 1963; Putaporntip et al. 2010; Warren and Wharton 1963
	Presbytis melalophos	Coatney et al. 1971; Contacos and Coatney 1963; Eyles 1963; Warren and Wharton 1963
P. simiovale	*Macaca sinica*	Coatney et al. 1971; Cogswell 2000; Escalante, Barrio, and Ayala 2005; Fooden 1994
P. simium	*Alouatta carya*	Duarte et al. 2008
	Alouatta guariba[14]	Coatney 1971; Coatney et al. 1971; Cogswell 2000; da Fonseca 1951; Deane 1964; Deane 1992; Duarte et al. 2008; Eyles 1963; Garnham 1963; Lourenço-de-Oliveira and Deane 1995; Stuart et al. 1998; Warren and Wharton 1963

	Brachyteles arachnoides	Coatney 1971; Coatney et al. 1971; Cogswell 2000; Deane 1992; Lourenço-de-Oliveira and Deane 1995
P. vivax	*Homo sapiens*	
	Pan troglodytes	Kaiser et al. 2010
P. cf. vivax	*Pan troglodytes*	Krief et al. 2010
Malariae Type		
P. brasilianum	*Alouatta belzebul*	Davies et al. 1991; Deane 1992; Lourenço-de-Oliveira and Deane 1995; Stuart et al. 1998
	Alouatta caraya	Deane 1992; Duarte et al. 2008[15]; Lourenço-de-Oliveira and Deane 1995; Stuart et al. 1998
	Alouatta guariba	Coatney et al. 1971; Cogswell 2000; Deane 1992; Duarte et al. 2008; Lourenço-de-Oliveira and Deane 1995; Stuart et al. 1998
	Alouatta palliata	Coatney et al. 1971; Cogswell 2000; Deane 1992; Eyles 1963; Lourenço-de-Oliveira and Deane 1995; Stuart et al. 1998
	Alouatta seniculus	Coatney et al. 1971; Cogswell 2000; Davies et al. 1991; de Thoisy et al. 2001; Fandeur et al. 2000; Lourenço-de-Oliveira and Deane 1995; Stuart et al. 1998; Volney et al. 2002
	Alouatta pigra[16]	Coatney et al. 1971; Cogswell 2000; Eyles 1963; Lourenço-de-Oliveira and Deane 1995; Porter, Johnson, and De Sousa 1966
	Alouatta spp.	Cogswell 2000; Warren and Wharton 1963
	Aotus vociferans	Collins et al. 1985a; Lourenço-de-Oliveira and Deane 1995
	Ateles belzebuth	Lourenço-de-Oliveira and Deane 1995
	Ateles chamek	Coatney et al. 1971
	Ateles fusciceps	Coatney et al. 1971; Cogswell 2000; Eyles 1963; Porter, Johnson, and De Sousa 1966
	Ateles geoffroyi[17]	Coatney et al. 1971; Cogswell 2000; Eyles 1963; Lourenço-de-Oliveira and Deane 1995; Porter, Johnson, and De Sousa 1966
	Ateles paniscus	Coatney et al. 1971; Cogswell 2000; Davies et al. 1991; Deane 1992; Dunn and Lambrecht 1963; Eyles 1963; Lourenço-de-Oliveira and Deane 1995
	Ateles spp.	Warren and Wharton 1963

	Brachyteles arachnoides	Coatney et al. 1971; Cogswell 2000; Deane 1992; Lourenço-de-Oliveira and Deane 1995
	Cacajao calvus[18]	Coatney 1971; Davies et al. 1991; Deane 1964; Deane 1992; Eyles 1963; Warren and Wharton 1963
	Callicebus brunneus	Davies et al. 1991; Deane 1992; Lourenço-de-Oliveira and Deane 1995
	Callicebus moloch	Coatney et al. 1971; Cogswell 2000; Davies et al. 1991; Deane 1992; Lourenço-de-Oliveira and Deane 1995
	Callicebus moloch complex	Davies et al. 1991
	Callicebus ornatus	Coatney et al. 1971; Lourenço-de-Oliveira and Deane 1995
	Callicebus torquatus	Coatney et al. 1971; Cogswell 2000; Davies et al. 1991; Deane 1992; Lourenço-de-Oliveira and Deane 1995
	Cebus albifrons	Coatney et al. 1971; Cogswell 2000; Davies et al. 1991; Eyles 1963
	Cebus apella	Coatney et al. 1971; Cogswell 2000; Davies et al. 1991; Deane 1992; Dunn and Lambrecht 1963; Eyles 1963; Lourenço-de-Oliveira and Deane 1995
	Cebus capucinus	Coatney et al. 1971; Cogswell 2000; Eyles 1963; Lourenço-de-Oliveira and Deane 1995; Porter, Johnson, and De Sousa 1966
	Cebus spp.	Warren and Wharton 1963
	Chiropotes albinasus	Davies et al. 1991; Deane 1992; Lourenço-de-Oliveira and Deane 1995
	Chiropotes chiropotes	Coatney et al. 1971; Cogswell 2000; Deane 1992
	Chiropotes satanas	Davies et al. 1991; Deane 1992; Lourenço-de-Oliveira and Deane 1995
	Lagothrix cana	Coatney et al. 1971; Cogswell 2000; Deane 1992
	Lagothrix poeppigii[19]	Coatney et al. 1971; Cogswell 2000; Deane 1992; Dunn and Lambrecht 1963; Eyles 1963
	Lagothrix lagotricha[20]	Coatney et al. 1971; Cogswell 2000; Davies et al. 1991; Deane 1992; Eyles 1963; Lourenço-de-Oliveira and Deane 1995

	Lagothrix sp. undet.	Warren and Wharton 1963
	Pithecia irrorata	Davies et al. 1991; Lourenço-de-Oliveira and Deane 1995
	Pithecia monachus	Deane 1992; Lourenço-de-Oliveira and Deane 1995
	Pithecia pithecia	Davies et al. 1991; Deane 1992; de Thoisy et al. 2001; Fandeur et al. 2000; Lourenço-de-Oliveira and Deane 1995; Volney et al. 2002
	Saguinus geoffroyi	Baerg 1971; Lourenço-de-Oliveira and Deane 1995
	Saguinus midas	Davies et al. 1991; Deane 1992; de Thoisy et al. 2001; Fandeur et al. 2000; Lourenço-de-Oliveira and Deane 1995; Volney et al. 2002
	Saimiri boliviensis	Coatney et al. 1971; Cogswell 2000; Davies et al. 1991; Deane 1992; Dunn and Lambrecht 1963; Eyles 1963; Lourenço-de-Oliveira and Deane 1995
	Saimiri sciureus	Coatney et al. 1971; Cogswell 2000; Davies et al. 1991; Deane 1992; Dunn and Lambrecht 1963; Eyles 1963; Lourenço-de-Oliveira and Deane 1995
	Saimiri ustus	Davies et al. 1991; Deane 1992; Lourenço-de-Oliveira and Deane 1995
	Saimiri sp. undet.	Garnham 1963; Warren and Wharton 1963
P. malariae	*Homo sapiens*	
	Pan troglodytes	Duval et al. 2010; Kaiser et al. 2010
P. cf. malariae	*Pan paniscus*	Krief et al. 2010
P. rodhaini	*Pan troglodytes*	Bray 1963; Coatney 1971; Coatney et al. 1971
Ovale-Type		
P. ovale	*Homo sapiens*	
	Pan troglodytes	Duval et al. 2009, 2010; Kaiser et al. 2010
Unclassified		
P. bucki	*Eulemur macaco*[21]	Gysin 1998
P. coulangesi	*Eulemur macaco*	Gysin 1998
P. eylesi	*Hylobates lar*	Coatney et al. 1971; Cogswell 2000; Warren, Coatney, and Skinner 1966

P. foleyi[22]	*Eulemur fulvus*[23]	Gysin 1998
	Lemurs	Cogswell 2000
P. georgesi	*Lophocebus albigena*[24]	Gysin 1998; Poirriez et al. 1995
	Cercocebus agilis	Gysin 1998; Poirriez et al. 1995
P. girardi	*Eulemur fulvus*	Coatney et al. 1971; Eyles 1963; Gysin 1998; Warren and Wharton 1963
P. jefferyi	*Hylobates lar*	Coatney, Orihel, and Warren 1969; Coatney et al. 1971; Cogswell 2000; Warren, Coatney, and Skinner 1966
P. lemuris[25]	Lemurs	Coatney et al. 1971
P. malagasi	*Propithecus verreauxi*	Duval et al. 2010
P. percygarnhami	*Eulemur macaco*	Gysin 1998
P. petersi	*Lophocebus albigena*	Gysin 1998; Poirriez, Dei-Cas, and Landau 1994
P. pitheci	*Pongo pygmaeus* s.l.[26]	Bray 1960; Coatney 1971; Coatney et al. 1971; Cogswell 2000; Garnham 1963; Peters et al. 1976; Wolfe et al. 2002
P. schwetzi	*Gorilla gorilla* s.l.	Bray 1963; Coatney 1971; Coatney et al. 1971; Cogswell 2000; Contacos and Coatney 1963; Warren and Wharton 1963
	Pan troglodytes	Bray 1958b; Bray 1963; Coatney 1971; Coatney et al. 1971; Cogswell 2000; Contacos and Coatney 1963; Warren and Wharton 1963
P. silvaticum	*Pongo pygmaeus* s.l.	Peters et al. 1976
P. uilenbergi	*Eulemur fulvus*	Gysin 1998
P. youngi	*Hylobates lar*	Coatney, Orihel, and Warren 1969; Coatney et al. 1971; Cogswell 2000; Eyles et al. 1964; Warren, Coatney, and Skinner 1966

[1]All gorillas were previously designated as a single species *Gorilla gorilla*, although with several subspecies. Contemporary systematics divide the eastern and western gorilla into two distinct species, *Gorilla beringei* and *Gorilla gorilla*. It is unclear from the studies that identified *P. reichenowi* and *P. schwetzi* in gorillas if it occurred in eastern or western gorillas.

[2]*Pan troglodytes* = former *Pan satyrus*

[3]*Macaca fascicularis* = former *Macaca irus, M. cynomolgus*

[4]Cogswell (2000) identifies *Macaca arctoides* (= speciosa) as a natural host of *P. cynomolgi*. The one case identified in the literature was by Coatney and associates (1971), who isolated *P. cynomolgi* in a zoo-housed, *Macaca arctoides*; however, they do not identify *Macaca arctoides* as a natural host.

[5]*Trachypithecus cristatus* = *Presbytis cristatus, Presbytis cristata*

[6]*Semnopithecus entellus* = *Presbytis entellus*

[7]Escalante and colleagues (2005) identify *Plasmodium fragile* as occurring in *Macaca mulatta* and *Presbytis* spp.; the literature suggests that they are not natural hosts, but infections have been induced experimentally.

[8]The literature sometimes identifies *Cercopithecus* spp. as natural hosts for *P. gonderi* (e.g., Escalante et al. 2005; Sullivan et al. 2002). These seem to be following a single identification in 1933 by Sinton and Mulligan in two vervet monkeys (*Cercopithecus aethiops = sabaeus*), which appears to have been isolated from a zoo environment (see Coatney et al. 1971). Given that no other cases have been identified, it is possible that either the primate was misidentified or the vervets acquired the parasite from mangabeys or drills in the zoo context, since they are susceptible to experimental infection.

[9]*Cerocebus agilis* = former *Cercocebus agilis galeritis.*

[10]*Lophocebus aterrimus = Cercocebus aterrimus* (also, aterrimus often mispelled as atterimus in the literature)

[11]*P. inui* = former *P. osmaniae* and former *P. shortii*

[12]*Macaca nigra* = former *Cynopithecus niger*

[13]*Trachypithecus obscurus = Presbytis obscura*

[14]*Alouatta guariba = Alouatta fusca*

[15]Duarte and associates do not distinguish between *P. simium/P. vivax* or *P. brasilianum/P.malariae* for *Alouatta carya* and *Alouatta guariba.*

[16]*Alouatta pigra = Alouatta villosa*

[17]*Ateles geoffroyi = Ateles grisestens, grisecens*

[18]Including *Cacajao rubicundus*

[19]Described as *Lagothrix infumata.* In the older literature, this designation has been used to refer to both *Lagothrix lagotricha* and *Lagothrix poeppigii* (see Fooden 1963). Dunn and Lambrecht (1963) identify the location as Loretto province in eastern Peru, so they are probably referring to *L. poeppigii.* Eyles (1963) and Cogwell's (2000) use of the term *Lagothrix lagotricha* is apparently following Dunn and Lambrecht's designation.

[20]*Lagothrix lagotricha = Lagothrix lagothricha*

[21]*Eulemur macaco = Lemur macaco*

[22]Little is known about the plasmodial parasites of prosmians. Cogswell (2000) suspects that *P. foleyi* is probably a hepatocystic parasite rather than a malarial parasite.

[23]*Eulemur fulvus = Lemur fulvus*

[24]*Lophocebus albigena = Cercocebus albigena*

[25]Some researchers consider *P. lemuris* and *P. girardi* to be conspecifics.

[26]All orangutans were previously designated as *Pongo pymaeus* (with two subspecies). In contemporary systematics, some are now differentiating orangutans as two species, the Bornean *Pongo pygmaeus* and the Sumatran *Pongo albelii.* The available studies of orangutan infection with *P. pitheci* and *P. silvaticum*, cannot be determined whether infections were documented in Sumatran or Bornean orangutans.

Appendix 2

Experimentally Induced[1] Plasmodium Cross-Infections into Novel Hosts

	Novel Hosts	References
Falciparum Type		
P. falciparum	*Alouatta palliata*[2]	Baerg and Young 1970; Coatney et al. 1971; Rossan and Baerg 1975; Stuart et al. 1998; Taliaferro and Taliaferro 1934
	Aotus azarae	Collins et al. 1986
	Aotus lemurinus	Collins et al. 1997a
	Aotus nancymaae[3]	Collins et al. 1988c
	Aotus trivirgatus	Coatney et al. 1971; Collins et al. 1973d; Collins et al. 1977; Collins et al. 1978; Collins et al. 1979c; Collins et al. 1979d; Collins et al. 1981a; Collins et al. 1982a; Collins et al. 1983a; Collins et al. 1986; Jervis et al. 1972; Miller 1969; O'Leary et al. 1972; Wellde et al. 1972
	Aotus vociferans	Collins et al. 1991
	Aotus "Peruvian"	Collins et al. 1986
	Ateles fusciceps	Baerg and Young 1970
	Cebus capucinus	Coatney et al. 1971
	Hylobates lar	Coatney et al. 1971; Gould and Cadigan 1966
	Macaca fascicularis	Coatney et al. 1971
	Macaca mulatta	Coatney et al. 1971
	Macaca nemestrina	Coatney et al. 1971
	Saimiri bolivensis	Campbell et al. 1986; Whitely et al. 1987
	Saimiri sciureus	Campbell et al. 1986; Coatney et al. 1971; Gysin, Hommel, and da Silva 1980

	Saguinus geoffroyi	Coatney et al. 1971; Porter and Young 1967
	Pan troglodytes	Bray 1958a; Taylor et al. 1985
Vivax Type		
P. coatneyi	*Macaca arctoides*[4]	Coatney et al. 1971; Eyles 1963
	Macaca nemestrina	Coatney et al. 1971
	Macaca mulatta	Aikawa et al. 1992; Coatney et al. 1971; Collins et al. 1967b; Collins et al. 2001; Collins et al. 2006; Contacos and Coatney 1963; Davison et al. 1998; Davison et al. 2000; Eyles 1963; Eyles et al. 1962; Eyles et al. 1963; Garnham 1963; Held and Contacos 1967; Maeno et al. 1993; Migot-Nabias et al. 1999; Miller and Fremount 1970; Miller, Chien, and Usami 1972; Millet et al. 1988; Moreno et al. 2007; Nakano et al. 1996; Sein et al. 1993; Smith et al. 1996; Tongren et al. 2000; Udomsangpetch et al. 1991; Udomsangpetch et al. 1993; Warren, Skinner, and Guinn 1966; Yang et al. 1999
	Macaca fuscata	Kawai et al. 1992; Kawai, Kano, and Suzuki 1995; Kawai et al. 2006; Matsumoto et al. 2000; Sugiyama et al. 2004
	Trachypithecus cristatus	Coatney et al. 1971; Eyles 1963; Palmieri et al. 1980
P. cynomolgi	*Aotus azarae*	Collins, Kinner, and Mehaffey 1985; Collins, Warren, and Galland 1999
	Aotus lemurinus	Collins, Warren, and Galland 1999; Collins et al. 2005
	Aotus nancymaae	Collins et al. 2005; Sullivan et al. 2006
	Aotus trivirgatus	Collins et al. 1975a; Collins et al. 1981a
	Cebus capucinus	Coatney et al. 1971
	Cebus sp. undet.	Contacos and Coatney 1963
	Cercopithecus aethiops	Coatney et al. 1971; Garnham 1963; Schmidt et al. 1977a
	Cercopithecus sp. undet.	Contacos and Coatney 1963; Garnham 1963
	Homo sapiens	Bennett and Warren 1965; Beye et al. 1961; Coatney 1971; Dissanaike 1965; Eyles, Coatney, and Getz 1960; Garnham, Molinari, and Shute 1962; Kuvin et al. 1962a; Kuvin et al. 1962b; Schmidt, Greenland, and Genther 1961; Tobie et al. 1961

	Papio papio	Coatney et al. 1971; Contacos and Coatney 1963; Garnham 1963
	Papio sp. undet.	Garnham 1963
	Saimiri boliviensis	Collins, Warren, and Galland 1999; Collins et al. 2005; Sullivan et al. 2006
P. fieldi	*Aotus lemurinus*	Sullivan et al. 1998
	Macaca fasicularis	Collins et al. 1992a; Held, Contacos, and Coatney 1967
	Macaca mulatta	Coatney et al. 1971; Collins and Contacos 1971; Collins et al. 1973e; Collins et al. 2006; Contacos and Coatney 1963; Eyles 1963; Garnham 1963
	Macaca radiata	Coatney et al. 1971
	Papio anubis[5]	Coatney et al. 1971
P. fragile	*Aotus lemurinus*	Collins et al. 2006
	Aotus nancymaae	Collins et al. 2006
	Aotus trivirgatus	Collins et al. 1974b; Collins et al. 1981a
	Aotus vociferans	Collins et al. 2006
	Macaca mulatta	Coatney et al. 1971; Collins et al. 1977; Collins et al. 1979; Collins et al. 2006; Fremount and Miller 1975; Puri and Dutta 1994
	Macaca fascicularis	Coatney et al. 1971; Collins et al. 1975b; Collins et al. 1992a; Collins et al. 2006
	Semnopithecus entellus	Puri and Dutta 1994
	Saimiri boliviensis	Collins et al. 1990b; Collins et al. 2006
P. gonderi	*Cercopithecus aethiops*	Coatney et al. 1971; Garnham 1963
	Cercopithecus sp. undet.	Contacos and Coatney 1963
	Macaca fasicularis	Collins et al. 1992a
	Macaca mulatta	Coatney et al. 1971; Collins and Contacos 1980; Contacos and Coatney 1963; Garnham 1963
	Macaca radiata	Coatney et al. 1971
	Papio anubis	Coatney et al. 1971
	Papio cynocephalus[6]	Coatney et al. 1971
	Papio ursinus	Contacos and Coatney 1963

	Papio sp. undet.	Garnham 1963
P. hylobati	*Aotus trivirgatus*	Collins et al. 1974a
	Macaca fascicularis	Coatney et al. 1971
	Macaca mulatta	Coatney et al. 1971; Collins et al. 1974a
	Macaca nemestrina	Coatney et al. 1971; Collins et al. 1974a
P. inui	*Aotus azarae*	Collins et al. 1988a
	Aotus trivirgatus	Collins et al. 1981b
	Cercopithecus mitis	Coatney et al. 1971
	Cercopithecus sp. undet.	Contacos and Coatney 1963
	Homo sapiens	Coatney 1971; Coatney et al. 1966; Coatney et al. 1971
	Saimiri boliviensis	Collins et al. 1988a; Sullivan et al. 2003
	Papio anubis	Contacos and Coatney 1963
P. knowlesi	*Aotus azarae*	Sullivan et al. 1996b
	Aotus trivigatus	Collins et al. 1981a
	Callithrix jacchus	Coatney et al. 1971; Garnham 1963
	Cebus sp. undet.	Coatney et al. 1971
	Cercocebus sp. undet.	Contacos and Coatney 1963; Garnham 1963
	Cercopithecus aethiops[7]	Coatney et al. 1971; Garnham 1963
	Cercopithecus atys	Coatney et al. 1971
	Cercopithecus cephus	Coatney et al. 1971
	Cercopithecus mona	Garnham 1963
	Cercopithecus sp. undet.	Contacos and Coatney 1963
	Homo sapiens	Chin et al. 1965; Chin et al. 1968; Coatney 1971
	Hylobates hoolock	Coatney et al. 1971
	Hylobates lar	Coatney et al. 1971
	Hylobates sp. undet.	Contacos and Coatney 1963; Garnham 1963
	Macaca arctoides	Coatney et al. 1971

	Macaca mulatta	Collins, Contacos, and Guinn 1967; Collins et al. 1967a; Collins et al. 1973b; Collins et al. 1973e; Collins et al. 2004b; Collins et al. 2006; Miller, Fremount, and Luse 1971; Millet et al. 1988; Pollet and Barr 1968; Rigdon and Stratman Thomas 1942; Rothe, Jacobus, and Walter 1969; Schmidt et al. 1977b; Warren, Skinner, and Guinn 1966
	Macaca radiata	Coatney et al. 1971; Garnham 1963
	Papio anubis	Coatney et al. 1971; Munyao et al. 2008; Ozwara et al. 2003
	Papio cynocephalus	Coatney et al. 1971
	Papio papio	Coatney et al. 1971; Garnham 1963
	Papio sp. undet.	Coatney et al. 1971; Contacos and Coatney 1963
	Saimiri boliviensis	Collins et al. 2004b; Sullivan et al. 1996b
	Saimiri sciureus	Coatney et al. 1971
	Semnopithecus entellus[8]	Coatney et al. 1971; Garnham 1963
	Trachypithecus cristatus	Coatney et al. 1971
P. simiovale	*Macaca mulatta*	Coatney et al. 1971; Collins and Contactos 1974; Collins, Contacos, and Jumper 1972; Collins et al. 2006
P. simium	*Alouatta palliata*	Coatney et al. 1971
	Alouatta pigra	Rossan, Baerg, and Young 1975
	Ateles paniscus	Coatney et al. 1971
	Aotus azarae	Collins et al. 1987b
	Aotus nancymaae	Collins et al. 1987b
	Aotus trivirgatus	Collins et al. 1973a; Collins, Contacos, and Jumper 1974; Collins et al. 1979b
	Aotus vociferans	Collins et al. 1987b
	Ateles fusciceps	Rossan, Baerg, and Young 1975
	Ateles geoffroyi	Rossan, Baerg, and Young 1975
	Ateles paniscus	Rossan, Baerg, and Young 1975
	Callithrix jacchus	Deane 1964; Rossan, Baerg, and Young 1975
	Cebus capucinus	Rossan, Baerg, and Young 1975
	Lagothrix lagotricha	Coatney et al. 1971; Rossan, Baerg, and Young 1975

	Saguinus geoffroyi	Rossan, Baerg, and Young 1975
	Saimiri oerstedii	Rossan, Baerg, and Young 1975
	Saimiri sciureus	Coatney 1971; Coatney et al. 1971; Collins, Contacos, and Jumper 1974; Deane 1964; Sterling, Aikawa, and Rabbege 1975
	Saimiri boliviensis	Collins et al. 1987b
P. vivax	*Aotus* "Bolivian"	Collins et al. 1983b
	Aotus azarae	Collins et al. 1985b; Collins et al. 1992b; Collins et al. 1994; Sullivan et al. 1999; Sullivan et al. 2001
	Aotus lemurinus	Collins et al. 1985b; Collins et al. 1992b; Collins et al. 1994; Collins et al. 2004a; Fremount and Rossan 1990; Sullivan et al. 1999; Sullivan et al. 2001
	Aotus nancymaae	Collins et al. 1988c; Collins et al. 1992b; Collins et al. 1994; Collins et al. 2004a; Sullivan et al. 1999; Sullivan et al. 2001
	Aotus trivirgatus	Baerg, Porter, and Young 1969; Coatney et al. 1971; Collins et al. 1973d; Collins et al. 1979d; Collins et al. 1980a; Collins et al. 1980b; Collins et al. 1981a; Collins et al. 1983b; Porter 1971
	Aotus vociferans	Collins et al. 1985b; Collins et al. 1987d; Sullivan et al. 1999; Sullivan et al. 2001
	Aotus hybrids	Collins et al. 1994; Sullivan et al. 1999; Sullivan et al. 2001
	Aotus sp. undet.	Sullivan et al. 1996a
	Ateles fusciceps	Baerg, Porter, and Young 1969; Coatney et al. 1971; Porter and Young 1970
	Ateles geoffroyi	Porter and Young 1970
	Cebus capucinus	Coatney et al. 1971
	Hylobates lar	Coatney et al. 1971; Cadigan, Ward, and Puhomchareon 1968
	Saguinus geoffroyi	Baerg, Porter, and Young 1969; Coatney et al. 1971
	Saimiri boliviensis[9]	Campbell et al. 1983; Collins et al. 1987c; Collins et al. 1988b; Collins et al. 1992b; Collins et al. 1994; Sullivan et al. 1999; Sullivan et al. 2001
	Saimiri sciureus[10]	Coatney et al. 1971; Collins et al. 1994; Fremount and Rossan 1990; Sullivan et al. 2001

	Saimiri sp. undet.	Deane, Neto, and Silveira 1966; Sullivan et al. 1996a
	Pan troglodytes	Bray 1957a; Coatney 1971; Coatney et al. 1971; Sullivan et al. 1996a
Malariae Type[11]		
P. brasilianum	*Homo sapiens*	Coatney 1971
	Aotus sp. undet.	Contacos and Coatney 1963
	Callithrix sp. undet.	Contacos and Coatney 1963
P. malariae	*Aotus azarae*	Collins et al. 1984; Collins et al. 1997b
	Aotus lemurinus	Collins et al. 1989a; Collins et al. 1997b
	Aotus nancymaae	Collins et al. 1988c; Collins et al. 1997b
	Aotus trivirgatus	Coatney et al. 1971; Geiman and Siddiqui 1969
	Saimiri boliviensis	Collins et al. 1997b
	Pan troglodytes	Bray 1957a; Coatney 1971; Collins et al. 1990a; Rodhain 1948
P. rodhaini	*Homo sapiens*	Coatney 1971
Ovale Type		
P. ovale	*Pan troglodytes*	Bray 1957b; Coatney et al. 1971; Morris et al. 1996
P. schwetzi[12]	*Homo sapiens*	Coatney 1971; Contacos et al. 1970
Unclassified		
P. eylesi	*Homo sapiens*	see Contactos et al. 1970
P. pitheci	*Pan troglodytes*	Coatney et al. 1971
P. silvaticum	*Pan troglodytes*	Gysin 1998
P. youngi	*Hylobates agilis*	Coatney et al. 1971
	Hylobates concolor	Coatney et al. 1971

[1]Includes transmitting by mosquito bite in a laboratory setting that is referred to in early prison studies as "naturally acquired."
[2]Rossan and Baerg describe species as *Alouatta villosa* = *Alouatta palliata*.
[3]*Aotus nancymaae* = *Aotus nancymai, Aotus nancymae*
[4]*Macaca arctoides* = former *Macaca speciosa*
[5]*Papio anubis* = *Papio doguera*
[6]*Papio cynocephalus* = *Papio jubilaeus*
[7]*Cercopithecus aethiops* = *Cecropithecus grisio viridis*
[8]Garnham refers to as *Semnopithecus entellus* "Assam hunuman (*sic*)."
[9]*Saimiri boliviensis* = *Saimiri sciureus boliviensis, Saimiri sciureus* "Peruvian"
[10]*Saimiri sciureus* includes *Saimiri sciureus* "Guyanan."
[11]Some researchers suspect that *P. brasilianum, P. rodhaini,* and *P. malariae.*
[12]The phylogenic placement of *P. schwetzi* has not been resolved, and some place it with the vivax-type malarias.

Appendix 3

Naturally Acquired Cross-Infections with Novel Malaria Parasites

	Novel Hosts	References
Falciparum Type		
	Alouatta caraya	Duarte et al. 2008
	Alouatta guariba	Duarte et al. 2008
Vivax Type		
P. cynomolgi[1]	*Homo sapiens*	Eyles, Coatney, and Getz 1960; Schmidt, Greenland, and Genther 1961; Schmidt et al. 1961
P. knowlesi	*Homo sapiens*	Bronner et al. 2009; Chin et al. 1965; Coatney 1971; Cox-Singh et al. 2008; Galinski and Barnwell 2009; Jongwuitwes 2004; Kantele et al. 2008; Kuo et al. 2009; Lee et al. 2009; Luchavez, Espino, and Curameng 2008; Putaporntip et al. 2009; Singh et al. 2004; van Hellemond et al. 2009; Vythilingham et al. 2008
P. inui	*Pongo pygmaeus* s.l.	Wolfe et al. 2002
P. simium	*Homo sapiens*	Deane, Deane, and Neto 1966
P. vivax[2]		
Malariae Type		
P. brasilianum	*Homo sapiens*	Contacos et al. 1963
P. malariae	*Pan troglodytes*	Hayakawa et al. 2009
Ovale Type		
P. ovale	*Pan troglodytes*	Duval et al. 2009

[1,2]Reid and associates (2006) identified naturally acquired infections of *P. cynomolgi* and *P. vivax* in orangutans. Singh and Divis (2009) have disputed this finding.

Appendix 4

Primate Species and All Infections with Plasmodium Parasites

Primate Species	Common Name	Natural Parasites	Cross-Infections
HUMAN			
Homo sapiens		*P. falciparum* *P. malariae* *P. ovale* *P. vivax*	*P. brasilianum* *P. cynomolgi* *P. eylesi* *P. inui* *P. knowlesi* *P. rodhaini* *P. schwetzi* *P. simium*
APE			
African			
Gorilla gorilla s.l.	Gorilla	*P. GorA* *P. GorB* *P. falciparum* *P. reichenowi* *P. schwetzi*	
Pan paniscus	Bonobo	*P. cf. falciparum* *P. cf. malariae*	
Pan troglodytes	Common chimpanzee	*P. billbrayi* *P. billcollensi* *P. falciparum* *P. gaboni* *P. malariae* *P. ovale* *P. reichenowi* *P. rodhaini* *P. schwetzi* *P. cf. vivax* *P. vivax*	*P. falciparum*[1] *P. malariae*[2] *P. ovale*[3] *P. pitheci* *P. silvaticum* *P. vivax*

Asian			
Hylobates agilis	Agile gibbon		*P. youngi*
Hylobates concolor	Black-crested gibbon		*P. youngi*
Hylobates lar	White-handed gibbon	*P. eylesi* *P. jeffreyi* *P. youngi*	*P. falciparum* *P. knowlesi* *P. vivax*
Hylobates hoolock	Hoolock gibbon		*P. knowlesi*
Hylobates moloch	Silvery gibbon	*P. hylobati*	
Hylobates sp. undet.	Gibbon		*P. knowlesi*
Pongo pygmaeus s.l.	Orangutan	*P. pitheci* *P. silvaticum*	*P. inui*
Old World Monkeys			
African			
Cercocebus agilis	Agile mangabey	*P. georgesi* *P. gonderi*	
Cercocebus atys	Sooty mangabey	*P. gonderi*	*P. knowlesi*
Cercocebus sp. undet.	Mangabey		*P. knowlesi*
Cercopithecus aethiops	Vervet monkey		*P. cynomolgi* *P. knowlesi*
Cercopithecus cephus	Moustached guenon		*P. knowlesi*
Cercopithecus mitis	Blue monkey		*P. inui*
Cercopithecus mona	Mona monkey		*P. knowlesi*
Cercopithecus sp. undet.		*P. gonderi*	*P. cynomolgi* *P. inui* *P. knowlesi*
Lophocebus albigena	Gray-cheeked mangabey	*P. georgesi*[4] *P. petersi*	
Lophocebus aterimus	Black mangabey	*P. gonderi*	
Mandrillus leucophaeus	Drill	*P. gonderi*	
Papio anubis	Olive baboon		*P. fieldi* *P. inui* *P. knowlesi*
Papio cynocephalus	Yellow baboon		*P. gonderi* *P. knowlesi*

Papio papio	Guinea baboon		*P. cynomolgi* *P. knowlesi*
Papio sp. undet.	Baboon		*P. cynomolgi* *P. knowlesi*
Papio ursinus	Chacma baboon		*P. gonderi*
Asian			
Macaca arctoides	Stump-tailed macaque	*P. cynomolgi*	*P. coatneyi* *P. knowlesi*
Macaca assamensis	Assamese macaque	*P. inui*	
Macaca cyclopis	Taiwan macaque Formosan macaque	*P. cynomolgi* *P. inui* *P. knowlesi*	
Macaca fascicularis	Crab-eating macaque Long-tailed macaque Kra monkey	*P. coatneyi* *P. cynomolgi* *P. fieldi* *P. inui* *P. knowlesi*	*P. falciparum* *P. fragile* *P. gonderi* *P. hylobati*
Macaca fuscata	Japanese macaque		*P. coatneyi*
Macaca mulatta	Rhesus macaque Rhesus monkey	*P. cynomolgi* *P. inui*	*P. coatneyi* *P. falciparum* *P. fragile* *P. fieldi* *P. gonderi* *P. hylobati* *P. knowlesi* *P. simiovale*
Macaca nemestrina	Pig-tailed macaque	*P. cynomolgi* *P. fieldi* *P. inui* *P. knowlesi*	*P. coatneyi* *P. falciparum* *P. hylobati*
Macaca nigra	Celebes-black macaque Sulawesi-crested macaque	*P. inui* *P. knowlesi*	
Macaca radiata	Bonnet macaque	*P. cynomolgi* *P. fragile* *P. inui*	*P. gonderi* *P. knowlesi*
Macaca sinica	Toque macaque	*P. cynomolgi* *P. fragile* *P. inui* *P. simiovale*	
Trachypithecus cristatus	Silver-leafed monkey Javan lutung	*P. cynomolgi* *P. inui*	*P. coatneyi* *P. knowlesi*

Semnopithecus entellus	Hanuman langur	*P. cynomolgi*	*P. fragile* *P. knowlesi*
Presbytis melalophos	Banded leaf monkey Sumatran surili	*P. knowlesi* *P. inui*	
Presbytis sp. undet.	Leaf monkey	*P. inui* *P. fragile*	
Trachypithecus obscurus	Dusky leaf monkey Spectacled langur	*P. inui*	
NEW WORLD MONKEYS			
Alouatta belzebul	Red-handed howler	*P. brasilianum*	
Alouatta caraya	Black howler	*P. brasilianum* *P. simium*	*P. falciparum*
Alouatta guariba	Brown howler	*P. brasilianum* *P. simium*	*P. falciparum*
Alouatta palliata	Mantled howler	*P. brasilianum*	*P. falciparum* *P. simium*
Alouatta seniculus	Red howler	*P. brasilianum*	
Alouatta pigra	Guatemalan black howler	*P. brasilianum*	*P. simium*
Aotus hybrid			*P. vivax*
Aotus azarae	Azara's night monkey		*P. cynomolgi* *P. falciparum* *P. inui* *P. knowlesi* *P. malariae* *P. simium* *P. vivax*
Aotus lemurinus	Colombian night monkey		*P. cynomolgi* *P. falciparum* *P. fieldi* *P. fragile* *P. malariae* *P. vivax*
Aotus nancymaae	Peruvian red-necked owl monkey		*P. cynomolgi* *P. falciparum* *P. fragile* *P. malariae* *P. simium* *P. vivax*
Aotus trivirgatus	Northern night monkey Douroucouli		*P. falciparum* *P. hylobati* *P. malariae* *P. simium*

Aotus vociferans	Spix's night monkey	*P. brasilianum*	*P. falciparum* *P. fragile* *P. simium* *P. vivax*
Aotus sp. undet.	Night monkey, owl monkey		*P. brasilianum*
Ateles belzebuth	Long-haired spider monkey	*P. brasilianum*	
Ateles chamek	Peruvian spider monkey	*P. brasilianum*	
Ateles fusciceps	Brown-headed spider monkey	*P. brasilianum*	*P. falciparum* *P. simium* *P. vivax*
Ateles geoffroyi	Black-handed spider monkey	*P. brasilianum*	*P. simium* *P. vivax*
Ateles paniscus	Black spider monkey	*P. brasilianum*	*P. simium*
Brachyteles arachnoides	Woolly spider monkey Southern Muriqui	*P. brasilianum* *P. simium*	
Cacajao calvus	Bald uakari	*P. brasilianum*	
Callicebus brunneus	Brown titi	*P. brasilianum*	
Callicebus moloch	Dusky titi	*P. brasilianum*	
Callicebus moloch complex[5]	Titi monkey	*P. brasilianum*	
Callicebus ornatus	Ornate titi	*P. brasilianum*	
Callicebus torquatus	Collared titi	*P. brasilianum*	
Callithrix jacchus	Common marmoset		*P. knowlesi* *P. simium*
Callithrix sp.undet.	Marmoset		*P. brasilianum*
Cebus albifrons	White-fronted capuchin	*P. brasilianum*	
Cebus apella	Tufted capuchin	*P. brasilianum*	
Cebus capucinus	White-faced capuchin	*P. brasilianum*	*P. falciparum* *P. cynomolgi* *P. simium* *P. vivax*

Cebus sp. undet.	Capuchin		*P. cynomolgi* *P. knowlesi*
Chiropotes albinasus	White-nosed saki	*P. brasilianum*	
Chiropotes chiropotes	Red-backed bearded saki	*P. brasilianum*	
Chiropotes satanas	Black-bearded saki	*P. brasilianum*	
Lagothrix cana	Peruvian woolly monkey	*P. brasilianum*	
Lagothrix poeppigii		*P. brasilianum*	
Lagothrix lagotricha	Silvery woolly monkey	*P. brasilianum*	*P. simium*
Pithecia irrorata	Bald-faced saki Tapajós saki	*P. brasilianum*	
Pithecia monachus	Monk saki	*P. brasilianum*	
Pithecia pithecia	White-faced saki	*P. brasilianum*	
Saguinus geoffroyi	Panamanian tamarin	*P. brasilianum*	*P. falciparum* *P. simium* *P. vivax*
Saguinus midas	Red-handed tamarin	*P. brasilianum*	
Saimiri boliviensis	Bolivian squirrel monkey	*P. brasilianum*	*P. cynomolgi* *P. falciparum* *P. fragile* *P. inui* *P. knowlesi* *P. malariae* *P. simium* *P. vivax*
Saimiri oerstedii	Central American squirrel monkey		*P. simium*
Saimiri sciureus	Common squirrel monkey	*P. brasilianum*	*P. falciparum* *P. knowlesi* *P. simium* *P. vivax*
Sairmiri ustus	Bare-eared squirrel monkey	*P. brasilianum*	
PROSIMIANS			
Eulemur fulvus	Brown lemur	*P. girardi* *P. uilenbergi*	
Eulemur macaco	Black lemur	*P. bucki* *P. coulangesi* *P. percygarnhami*	

Propithecus verreauxi	White sifaka	*P. malagasi*	
Lemurs		*P. foleyi* *P. girardi* *P. lemuris*	

[1,2,3]*P. falciparum, P. malariae,* and *P. ovale* have been experimentally induced in chimpanzees but recently also discovered in wild chimpanzees.
[4]*P. georgesi* is newly identified.
[5]Moloch complex = *Callicebus baptista, C. bernhardi, C. brunneus, C. cinerascens, C. hoffmannsi,* and *C. moloch.*

References

Abegg, C., and B. Thierry. 2002. Macaque evolution and dispersal in insular Southeast Asia. *Biological Journal of the Linnean Society* 75(4): 555–576.
Adam, I., S. Babkier, A. A. Mohmmed, M. M. Salih, M. H. Prins, and Z. M. Zaki. 2007. ABO blood group system and placental malaria in an area of unstable malaria transmission in Eastern Sudan. *Malaria Journal* 6: 110 (August 10, 2007), *www.malariajournal.com/content/6/1/110* (accessed December 16, 2008).
Adeola, F. O. 2004. Boon or bane? The environmental and health impacts of persistant organic pollutants. *Human Ecology Review* 11(1): 27–35.
Aikawa, M., A. Brown, C. D. Smigh, T. Tegoshi, R. J. Howard, T. H. Hasler, Y. Ito, G. Perry, W. E. Collins, and K. Webster. 1992. A primate model for human cerebral malaria: *Plasmodium coatneyi*-infected Rhesus monkeys. *American Journal of Tropical Medicine and Hygiene* 46(4): 391–397.
Ankel-Sions, F., J. G. Fleagle, and P. S. Chatrath. 1998. Femoral anatomy of *Aegyptopithecus zeuxis*, an early Oligocene anthropoid. *American Journal of Physical Anthropology* 106(4): 413–424.
Alden, D., and J. C. Miller. 1987. Out of Africa: The slave trade and the transmission of smallpox to Brazil, 1560–1831. *Journal of Interdisciplinary History* 18: 195–224.
Algren, N. 1951. *Chicago: A city on the make*. Chicago: University of Chicago Press.
Allen, S. J., A. O'Donnell, N. D. E. Alexander, C. S. Mgone, T. E. A. Peto, J. B. Clegg, M. P. Alpers, and D. J. Weatherall. 1999. Prevention of cerebral malaria in children in Papua New Guinea by Southeast Asian ovalocytosis band 3. *American Journal of Tropical Medicine and Hygiene* 60(6): 1056–1060.
Allison, A. C. 2009. Genetic control of resistance to human malaria. *Curent Opinion in Immunology* 21: 499–505.
———. 1954. Protection afforded by sickle-cell trait against subteritian malarial infection. *British Medical Journal* 1(4857): 290–294.
Allison, A. C., and D. F. Clyde. 1961. Malaria in African children with deficient erythrocyte glucose-6-phosphate dehydrogenase. *British Medical Journal* 1(5236): 1346–1349.
Alving, A. S., B. Craige, T. N. Pullman, C. Merrill Whorton, R. Jones, and L. Eichelberger. 1948. Procedures used at Stateville Penitentiary for the testing of potential antimalarial agents. *Journal of Clinical Investigation* 27: 2–5.

Anderson, R. A., B. G. J. Knols, and J. C. Koella. 2000. *Plasmodium faliciparum* sporozoites increase feeding-associated mortality of their mosquito hosts *Anopheles gambiae* s.1. *Parasitology* 120(4): 329–333.

Anderson, R. A., J. C. Koella, and H. Hurd. 1999. The effect of *Plasmodium yoelli nigeriensis* infection on the feeding persistence of *Anopheles stephensis* Liston throughout the sporogenic cycle. *Proceedings of the Royal Society of London: Biological Sciences* 266(1430): 1729–1733.

Andrews, P., and J. Kelley. 2007. Middle Miocene dispersals of apes. *Folia Primatologia* 78: 328–343.

Anthony, N. M., M. Johnson-Bawe, K. Jeffery, S. L. Clifford, K. A. Abernethy, C. E. Tutin, S. A. Lahm, L. J. T. White, J. F. Utley, E. J. Wickings, and M. W. Bruford. 2007. The role of Pleistocene refugia and rivers in shaping gorilla genetic diversity in central Africa. *Proceedings of the National Academy of Sciences, United States of America* 104(51): 20432–20436.

Anonymous. 1961. Trouble in the monkey business. *Science News-Letter* 79(6): 90–91.

Antón, S. C., W. R. Leonard, and M. L. Robertson. 2002. An ecomorphological model of the initial hominid dispersal from Africa. *Journal of Human Evolution* 43(6): 773–785.

Antón, S. 2003. Natural history of *Homo erectus*. *American Journal of Physical Anthropology* 122(S37): 126–170.

Anstey, N. M., B. Russell, T. W. Yeo, and R. N. Price. 2009. The pathphysiology of *vivax* malaria. *Trends in Parasitology* 25(5): 220–227.

Arens, R. 1972. Using human guinea pigs in syphilis study. *The New York Times*, August 17, 1972, p. 34.

Argue, D., D. Donlon, C. Groves, and R. Wright. 2006. *Homo floresiensis:* Microcephalic, pygmoid, *Australopithecus* or *Homo? Journal of Human Evolution* 51(4): 360–374.

Armour-Chelu, M., P. Andrews, and R. Bernor. 2005. Further observations on the primate community at Rudabánya II (late Miocene, early Vallesian age), Hungary. *Journal of Human Evolution* 49(1): 88–98.

Athreya, B. H., and L. L. Coriell. 1967. Relation of blood groups to infection. I. A survey and review of the data suggesting possible relationship between malaria and blood groups. *American Journal of Epidemiology* 88(2): 292–304.

Augustin, C. B., P. R. F. Barnes, J. M. Barnola, M. Bigler, E. Castellano, O. Cattani, J. Chappellaz, D. Dahl-Jensen, B. Delmonte, G. Dreyfus, G. Durand, S. Falourd, H. Fischer, J. Flückiger, M. E. Hansson, P. Huybrechts, G. Jugie, S. J. Johnsen, J. Jouzel, P. Kaufmann, J. Kipfstuhl, F. Lambert, V. Y. Lipenkov, G. C. Littot, A. Longinelli, R. Lorrain, V. Maggi, V. Masson-Delmotte, H. Miller, R. Mulvaney, J. Oerlemans, H. Oerter, G. Orombelli, F. Parrenin, D. A. Peel, J.-R. Petit, D. Raynaud, C. Ritz, U. Ruth, J. Schwander, U. Siegenthaler, R. Souchez, B. Stauffer, J. P. Steffensen, B. Stenni, T. F. Stocker, I. E. Tabacco, R. Udisti, R. S. W. van de Wal, M. van den Broeke, J. Weiss, F. Wilhelms, J.-G. Winther, E. W. Wolff, and M. Zucchelli. 2004. Eight glacial cycles from an Antactic ice core. *Nature* 429: 623–628.

Ayala, F. J., A. A. Escalante, and S. M. Rich. 1999. Evolution of *Plasmodium* and the recent origin of the world's populations of *Plasmodium falciparum*. *Parassitologia* 41(1–3): 55–68.

Ayala, F. J., A. Escalante, A. A. Lal, and S. M. Rich. 1998. Evolutionary relationships of human malaria parasites. In *Malaria: Parasites, Pathogenesis, and Protection,*

Irwin W. Sherman, Ed., 285–300. Washington, D.C.: American Society for Microbiology Press.

Baerg, D. C. 1971. A naturally acquired infection of *Plasmodium brasilianum* in the marmoset, *Saguinus geoffroyi*. *Journal of Parasitology* 57(1): 8.

Baerg, D. C., J. A. Porter Jr., and M. D. Young. 1969. Sporozoite transmission of *Plasmodium vivax* to Panamanian primates. *American Journal of Tropical Medicine and Hygiene* 18(3): 346–350.

Baerg, D. C., and M. D. Young. 1970. *Plasmodium falciparum* infection induced in the black spider monkey, *Ateles fusciceps*, and the black howler monkey, *Alouatta villosa*. *Transactions of the Royal Society of Tropical Medicine and Hygiene* 64(1): 193–194.

Bailey, S. E., T. D. Weaver, and J.-J. Hublin. 2009. Who made the Aurignacian and other early Upper Paleolithic industries? *Journal of Human Evoution* 57(1): 11–26.

Baldwin, J. D., and J. Baldwin. 1976. Primate populations in Chiriquí, Panama. In *Neotropical primates: Field studies and conservation*, Eds. R. W. Thorington and P. B. Heltne, 20–31. Washington D.C.: National Academy of Sciences.

Balée, W. 1994. *Footprints of the forest: Ka'apor ethnobotany—The historical ecology of plant utilization by an Amazonian people*. New York: Columbia University Press.

———. 2006. The research program in historical ecology. *Annual Review of Anthropology* 35: 75–98.

Balée, W., and C. Erickson, Eds. 2006. *Time and complexity in historical ecology: Studies from the neotropical lowlands*. New York: Columbia University Press.

Barker, G., H. Barton, M. Bird, P. Daly, I. Datan, A. Dykes, L. Farr, D. Gilbertson, B. Harrisson, C. Hunt, T. Higham, L. Kealhofer, J. Krigbaum, H. Lewis, S. McLaren, V. Paz, A. Pike, P. Piper, B. Pyatt, R. Rabett, T. Reynolds, J. Rose, G. Rushworth, M. Stephens, C. Stringer, J. Thompson, and C. Turney. 2007. The "human revolution" in lowland tropical Southeast Asia: The antiquity and behavior of anatomically modern humans at Niah Cave (Sarawak, Borneo). *Journal of Human Evolution* 52(3): 243–261.

Bechtel, L. K., and C. P. Holstege. 2007. Criminal poisoning: Drug facilitated sexual assault. *Emergency Medicine Clinics of North America* 25(2) (May 2007), *http://das/journal/view/0/N/19495174?ja=581652&PAGE=1.html&issn=0733-8627&source=*(accessed December 7, 2009).

Benefit, B. R., and M. L. McCrossin. 1997. Earliest known Old World monkey skull. *Nature* 388(6640): 327–328.

Benefit, B. R., and M. L. McCrossin. 2002. Old World monkeys: The Miocene emergences. In *The Primate Fossil Record*, W. Hartwig, Ed., 241–253. Cambridge: Cambridge University Press.

Bennett, G. F., and M. Warren. 1965. Transmission of a new strain of *Plasmodium cynomolgi* to man. *Journal of Parasitology* 51(1): 79–80.

Benton, M. J., and P. C. J. Donoghue. 2006. Paleontological evidence to date the tree of life. *Molecular Biology and Evolution* 24(1): 26–53.

Bergsten, J. 2005. A review of long-branch attraction. *Cladistics* 21(2): 163–193.

Beye, H. K., M. E. Getz, G. R. Coatney, H. A. Elder, and D. E. Eyles. 1961. Simian malaria in man. *American Journal of Tropical Medicine and Hygiene* 10(3): 311–316.

Blatteis, C. M. 2003. Fever: Pathological or physiological, injurious or beneficial? *Journal of Thermal Biology* 28(1): 1–13.

Bloch, A. B., W. A. Orenstein, H. C. Stetler, S. G. Wassilak, R. W. Amler, K. J. Bart, C. D. Kirby, and A. R. Hinman. 1985. Health impact of measles vaccination in the United States. *Pediatrics* 76(4): 524–532.

Blum, D. 1994. *The monkey wars.* New York: Oxford University Press.

———. 2002. *Love at Goon Park: Harry Harlow and the science of affection.* Cambridge: Perseus Publishing.

Blume, S., and I. Geesink. 2000. A brief history of polio vaccines. *Science* 288(5471): 1593–1594.

Boaz, N. T., R. L. Ciochon, Q. Xu, and J. Liu. 2004. Mapping and taphonomic analysis of the *Homo erectus* loci at Locality 1 Zhoukoudian, China. *Journal of Human Evolution* 46(5): 519–549.

Bosselman, F. P., and D. A. Tarlock. 1993. The influence of ecological science on American law: An introduction. *Kent Law Review* 69: 847–873.

Botting, J. 2002. The history of thalidomide. *Drug News and Perspectives* 15(9): 604.

Boyd, M. F., Ed. 1949. *Malariology: A Comprehensive Survey of All Aspects of this Group of Diseases from a Global Standpoint.* Philadelphia: W. B. Saunders.

Braks, M. A. H., R. A. Anderson, and B. G. J. Knols. 1999. Infochemicals in mosquito host selection: Human skin microflora and *Plasmodium* parasites. *Parasitology Today* 15(10): 409–413.

Bray, R. S. 1957a. Studies on malaria in chimpanzees. II: *Plasmodium vivax. American Journal of Tropical Medicine and Hygiene* 6(3): 514–520.

———. 1957b. Studies on malaria in chimpanzees: IV: *Plasmodium ovale. American Journal of Tropical Medicine and Hygiene* 6(4): 638–645.

———. 1958a. Studies on malaria in chimpanzees. VI: *Laverania falciparum. American Journal of Tropical Medicine and Hygiene* 7(1): 20–24.

———. 1958b. Studies on malaria in chimpanzees. V: The sporogonous cycle and mosquito transmission of *Plasmodium vivax schwetzi. American Journal of Tropical Medicine and Hygiene* 44(1): 46–51.

———. 1963. The malaria parasites of anthropoid apes. *Journal of Parasitology* 49(6): 888–891.

Bronner, U., P. C. S. Divis, A. Färnert, and B. Singh. 2009. Swedish traveler with *Plasmodium knowlesi* malaria after visiting Malaysian Borneo. *Malaria Journal* 8: 15 (January 16), *www.malariajournal.com/content/8/1/15* (accessed October 16, 2009).

Brooks, D. R., and D. A. McLennan. 1992. The evolutionary origin of *Plasmodium falciparum. Journal of Parasitology* 78(3): 564–566.

Brown, P., T. Sutikna, M. J. Morwood, R. P. Soejono, Jatmiko, E. Wayhu Saptomo, and Rokus Awe Due. 2004. A new small-bodied hominin from the Late Pleistocene of Flores, Indonesia. *Nature* 431(7012): 1055–1061.

Bruce-Chwatt, L. J. 1965. Paleogenesis and paleo-epidemiology of primate malaria. *Bulletin of the World Health Organization* 32: 363–387, *http://whqlibdoc.who.int/bulletin/1965/Vol32/Vol32-No3/bulletin_1965_32(3)_363-387.pdf* (accessed November 27, 2008).

Bruce-Chwatt, L. J., and J. de Zulueta. 1980. *The rise and fall of malaria in Europe: A historico-epidemiological study.* Oxford: Oxford University Press.

Caccone, A., and J. R. Powell. 1989. DNA divergence among hominoids. *Evolution* 43(5): 925–942.

Cadigan, F. C., Jr., R. A. Ward, and S. Puhomchareon. 1968. Transient infection of the gibbon with *Plasmodium vivax* malaria. *Transactions of the Royal Society of Tropical Medicine and Hygiene* 62(2): 295–296.

Campbell, C. C., W. E. Collins, W. Chin, J. M. Roberts, and J. R. Broderson. 1983. Studies of the Sal I strain of *Plasmodium vivax* in the Squirrel monkey (*Saimiri sciureus*). *Journal of Parasitology* 69(3): 598–601.

Campbell, C. C., W. E. Collins, W. K. Milhaus, J. M. Roberts, and A. Armstead. 1986. Adaptation of the Indochina I/CDC strain of *Plasmodium falciparum* to the Squirrel monkey (*Saimiri sciureus*). *American Journal of Tropical Medicine and Hygiene* 35(3): 472–475.

Capellini, M. D., and G. Fiorelli. 2008. Glucose-6-phosphate dehydrogenase deficiency. *Lancet* 371(9606): 64–74.

Carfaro, P. 2001. Thoreau, Leopold, and Carson: Toward and environmental virtue ethics. *Environmental Ethics* 22: 4–17.

Carlton, J. M., J. H. Adams, J. C. Silva, S. L. Bidwell, H. Lorenzi, E. Caler, J. Crabtree, S. V. Angiuoli, E. F. Merino, P. Amedeo, Q. Cheng, R. M. R. Coulson, B. S. Crabb, H. A. del Portillo, K. Essien, T. V. Feldblyum, C. Fernandez-Becerra, P. R. Gilson, A. H. Gueye, X. Guo, S. Kang'a, T. W. A. Kooij, M. Korsinczky, E. V. S. Meyer, V. Nene, I. Paulsen, O. White, S. A. Ralph, Q. Ren, T. J. Sargeant, S. L. Salzberg, C. J. Stoeckert, S. A. Sullivan, M. M. Yamamoto, S. L. Hoffman, J. R. Wortman, M. J. Gardner, M. R. Galinski, J. W. Barnwell, and C. M. Fraser-Liggett. 2008. Comparative genomics of the neglected human malaria parasite *Plasmodium vivax*. *Nature* 455: 757–763.

Carmichael, A. G. 1993. Diseases of the Renaissance and early modern Europe. In *The Cambridge World History of Human Disease*, Kenneth F. Kiple, Ed., 279–287. Cambridge: Cambridge University Press.

Carroll, L. 1865. *Alice's adventures in wonderland*. New York: Appleton.

Carson, R. 1962. *Silent spring*. Boston: Houghton Mifflin.

Carter, R. 2003. Speculations on the origins of *Plasmodium vivax* malaria. *Trends in Parasitology* 19(5): 214–219.

Carter, R., and K. N. Mendis. 2002. Evolutionary and historical aspects of the burden of malaria. *Clinical Microbiology Reviews* 15: 564–594.

Cavalli-Sforza, L. L., and M. W. Feldman. 2003. The application of molecular genetic approaches to the study of human evolution. *Nature Genetics* 33: 266–275.

Cavasini, C. E., L. C. de Mattos, Á. A. R. D'Almeida Couto, V. S.C. D'Almeida Couto, Y. Gollino, L. J. Moretti, C. R. Bonini-Domingos, A. R. B. Rossit, L. Castilho, and R. L. D. Machado. 2007. Duffy blood group gene polymorphisms among malaria *vivax* patients in four areas of the Brazilian Amazon Region. *Malaria Journal* 6: 167, *www.malariajournal.com/content/6/1/167* (accessed November 23, 2008).

Cavasini, M. T. V., W. L. Ribeiro, F. Kawamoto, and M. U. Ferriera. 2000. How prevalent is *Plasmodium malariae* in Rondonia, Western Brazilian Amazon? *Revista da Sociedade Brasleira de Medicina Tropical* 33: 489–492.

Centers for Disease Control and Prevention (CDC). 2008. Syphilis-CDC Fact Sheet, *www.cdc.gov/std/syphilis/STDFact-Syphilis.htm* (accessed August 4, 2010).

———. 2006. Malaria: Schema of the life cycle of malaria, *www.cdc.gov/malaria/biology/life_cycle.htm* (accessed December 18, 2009).

———. 1998. Fatal cercopithecine herpesvirus 1 (B virus) infection following a mucocutaneous exposure and interim recommendations for worker protection. *Morbidity and Mortality Weekly Report* 47: 1073–1076, 1083.

———. 1997. Nonhuman primate spumavirus infections among persons with occupational exposure—United States, 1996. *Morbidity and Mortality Weekly Report* 46: 129–131.

———. 1987a. Epidemiologic notes and reports B-virus infection in humans: Pensacola, Florida. *Mortality and Morbidity Weekly Report* 36: 289–290, 295–296.

———. 1987b. Guidelines for prevention of *Herpesvirus simiae* (B virus) infection in monkey handlers. *Morbidity and Mortality Weekly Report* 36: 680–682, 687–689.

Chaimanee, Y., D. Jolly, M. Benammi, P. Tafforeau, D. Duzer, I. Moussa, and J.-J. Jaeger. 2003. A middle Miocene hominoid from Thailand and orangutan origins. *Nature* 422(6927): 61–65.

Chaimanee, Y., V. Suteethorn, P. Jintasakul, C. Vidthayanon, B. Marandat, and J.-J. Jaeger. 2004. A new orangutan relative from the late Miocene of Thailand. *Nature* 427(6973): 439–441.

Chaimanee, Y., C. Yamee, P. Tian, O. Chavasseau, and J.-J. Jaeger. 2008. First middle Miocene sivaladapid primate from Thailand. *Journal of Human Evolution* 54: 434–443.

Chaimanee, Y., C. Yamee, P. Tian, K. Khaowiset, B. Marandat, P. Tafforeau, C. Nemoz, and J.-J. Jaeger. 2006. *Khoratpithecus piriyai*, a late Miocene hominoid of Thailand. *American Journal of Physical Anthropology* 131(3): 311–323.

Chandre, F., F. Darrier, L. Manga, M. Akogbeto, O. Faye, J. Mouchet, and P. Guillet. 1999. Status of pyrethroid resistance in *Anopheles gambiae* sensu lato. *Bulletin of the World Health Organization* 77(3): 230–234.

Chakravorty, R. C. 1993. Diseases of antiquity in south Asia. In *The Cambridge World History of Human Disease*, Kenneth F. Kiple, Ed., 408–413. Cambridge: Cambridge University Press.

Chen, F.-C., and W.-H. Li. 2001. Genomic divergences between humans and other hominoids and the effective population size of the common ancestor of humans and chimpanzees. *American Journal of Human Genetics* 68(2): 444–456.

Chernin, E. 1984. The malariatherapy of neurosyphilis. *Journal of Parasitology* 70(5): 611–617.

Chimpanzee Sequencing and Analysis Consortium. 2005. Initial sequence of the chimpanzee genome and comparison with the human genome. *Nature* 437: 69–87.

Chin, W., and P. G. Contacos. 1966. A recently isolated West African strain of *Plasmodium ovale. American Journal of Tropical Medicine and Hygiene* 15(1): 1–2.

Chin, W., P. G. Contacos, and J. N. Buxbaum. 1966. The transmission of a West African strain of *Plasmodium ovale* by *Anopheles freeborni* and *Anopheles maculatus. American Journal of Tropical Medicine and Hygiene* 15(5): 690–693.

Chin, W., P. G. Contacos, G. R. Coatney, and H. R. Kimball. 1965. A naturally acquired Quotidian-type malaria in man transferable to monkeys. *Science* 149(3686): 865.

Chin, W., P. G. Contacos, and G. R. Coatney. 1966. The evaluation of quinine sulfate and diaminodiphenyl sulfone (DDS) as suppressive agents against sporozoite-induced Chesson strain *vivax* malaria. *American Journal of Tropical Medicine and Hygiene* 15(6): 830–832.

Chin, W., and G. R. Coatney. 1971. Relapse activity in sporozoite-induced infections with a West African strain of *Plasmodium ovale. American Journal of Tropical Medicine and Hygiene* 20(6): 825–827.

Chin, W., G. R. Coatney, and H. K. King. 1967. An evaluation of CI-564 against blood-induced chlorguanide-resistant strains of *vivax* malaria. *American Journal of Tropical Medicine and Hygiene* 16(1):13–14.

Chin, W., P. G. Contacos, G. R. Coatney, M. H. Jeter, and E. Alpert. 1967. Evaluation of CI-564, a 1:1 mixture of cycloguanil pamoate (CI-501) and 4,4'-diacetylaminodiphenylsulfone (CI-556), against multiresistant *falciparum* malarias. *American Journal of Tropical Medicine and Hygiene* 16(5): 580–584.

Chin, W., P. G. Contacos, G. R. Coatney, and H. K. King. 1966. The evaluation of sulfonamides, alone or in combination with pyrimethamine, in the treatment of multi-drug resistant *falciparum* malaria. *American Journal of Tropical Medicine and Hygiene* 15(6): 823–829.

Chin, W., P. G. Contacos, W. E. Collins, M. H. Jeter, and E. Alpert. 1968. Experimental mosquito-transmission of *Plasmodium knowlesi* to man and monkey. *American Journal of Tropical Medicine and Hygiene* 17(3): 355–358.

Chou, H.-H., H. Takematsu, S. Diaz, J. Iber, E. Nickerson, K. L. Wright, E. A. Muchmore, D. L. Nelson, S. T. Warren, and A. Varki. 1998. A mutation in human CMP-sialic acid hydroxylase occurred after the *Homo-Pan* divergence. *Proceedings of the National Academy of Science, United States of America* 95(20): 11751–11756.

Clark, A. G. 2002. Malaria variorum. *Nature* 418: 283–285.

Clark, H. C., and L. H. Dunn. 1931. Experimental efforts to transfer monkey malaria to man. *Journal of Tropical Medicine and Hygiene* s1-11(1): 1–10.

Cleary, D. 2001. Towards an environmental history of the Amazon: From prehistory to the 19th century. *Latin American Research Review* 36: 64–91.

Clyde, D. F., R. M. Miller, A. R. Schwartz, and M. M. Levine. 1971. Treatment of *falciparum* malaria with sulfalene and trimethoprim. *American Journal of Tropical Medicine and Hygiene* 20(6): 804–810.

Coatney, G. R. 1968. Simian malarias in man: Facts, implications, and predictions. *American Journal of Tropical Medicine and Hygiene* 17(2): 147–155.

———. 1971. The simian malarias: Zoonoses, anthroponoses or both? *American Journal of Tropical Medicine and Hygiene* 20(6): 795–803.

Coatney, G. R., W. Chin, P. G. Contacos, and H. K. King. 1966. *Plasmodium inui*, a Quartan-type malaria parasite of Old World monkeys transmissible to man. *Journal of Parasitology* 52(4): 660–663.

Coatney, R. G., P. G. Contacos, and J. S. Lunn. 1964. Further observations of the antimalarial activity of CI-501 (Camolar®) against the Chesson strain of *vivax* malaria. *American Journal of Tropical Medicine and Hygiene* 13(3): 383–385.

Coatney, R. G., P. G. Contacos, J. S. Lunn, J. W. Kilpatrick, and H. A. Elder. 1963. The effect of a repository preparation of the dihydrotriazine metabolite of chlorguanide, CI-501, against the Chesson strain of *Plasmodium vivax* in man. *American Journal of Tropical Medicine and Hygiene* 12(4): 504–508.

Coatney, G. R., H. A. Elder, P. G. Contacos, M. E. Getz, R. Greenland, R. N. Rossan, and L. H. Schmidt. 1961. Transmission of the M strain of *Plasmodium cynomolgi* to man. *American Journal of Tropical Medicine and Hygiene* 10(5): 673–678.

Coatney, G. R., A. V. Myatt, T. Hernandez, G. M. Jefferey, and W. Clark Cooper. 1953. Studies in human malaria. XXXII: The protective and therapeutic effects of pyrimethamine (daraprim) against Chesson strain of *vivax* malaria. *American Journal of Tropical Medicine and Hygiene* 2(5): 777–787.

Coatney, G. R., Thomas C. Orihel, and McWilson Warren. 1969. *Plasmodium jeffreryi*: A rediscription. *Journal of Parasitology* 55(6): 1235–1239.

Coatney, G. R., W. E. Collins, M. Warren, and P. G. Contacos. 1971. *The Primate Malarias*. Bethesda: U.S. Department of Health, Education and Welfare. Also available online at *www.dpd.cdc.gov/DPDx/HTML/PDF_Files/Primate%20 Malarias%20Chapters/front_matter.pdf* and as a CD-ROM (2003).

Coggeshall, L. T. 1941. Infection of *Anopheles quadrimaculatus* with *Plasmodium cynomolgi*, a monkey malaria parasite, and with *Plasmodiuml Lophurae*, an avian malaria parasite. *American Journal of Tropical Medicine and Hygiene* s1-21(4): 515–530.

Cogswell, F. B. 2000. Malaria and piroplasms of non-human primates. In *Companion and Exotic Animal Parasitology*, D. D. Bowman, Ed., International Veterinary Information Service. *www.ivis.org/advances/Parasit_Bowman/cogswell_primate/ ivis.pdf* (accessed November 21, 2008).

Cohen, J. 2009. Key malaria parasite likely evolved from chimp version. *Science* 325(5941): 633.

Coimbra, C. E. A. 1988. Human settlements, demographic pattern, and epidemology in lowland Amazonia: The case of chagas's disease. *American Anthropologist* 90: 82–97.

Collins, W. E., C. C. Campbell, J. C. Skinner, W. Chin, P. Nguyen-Dinh, and A. Y. Huong. 1983a. Studies on the Indochina I/CDC strain of *Plasmodium falciparum* in Colombian and Bolivian *Aotus* monkeys and different anophelines. *Journal of Parasitology* 69(1): 186–190.

Collins, W. E., W. Chin, and J. C. Skinner. 1979. *Plasmodium fragile* and *Macaca mulatta* monkeys as a model system for the study of malaria vaccines. *American Journal of Tropical Medicine and Hygiene* 28(6): 948–954.

Collins, W. E., W. Chin, M. Warren, A. Y. Huong, G. M. Jeffery, and J. C. Skinner. 1982a. Observations on two strains of *Plasmodium falciparum* from Haiti in *Aotus* monkeys. *Journal of Parasitology* 68(4): 657–667.

Collins, W. E., and P. G. Contacos. 1971. Observations on the relapse activity of *Plasmodium fieldi* in the Rhesus monkey. *Journal of Parasitology* 57(1): 29–32.

———. 1974. Observations on the relapse activity of *Plasmodium simiovale* in the Rhesus monkey. *Journal of Parasitology* 60(2): 343.

———. 1979. Infection and transmission studies with *Plasmodium simiovale* in the *Macaca mulatta* monkey. *Journal of Parasitology* 65(4): 609–612.

———. 1980. Infection and transmission studies with *Plasmodium gonderi* in the *Macaca mulatta* monkey. *Journal of Parasitology* 66(6): 998–1002.

Collins, W. E., P. G. Contacos, P. C. C. Garnham, McWilson Warren, and Jimmie C. Skinner. 1972. *Plasmodium hylobati:* A malaria parasite of the gibbon. *Journal of Parasitology* 58(1): 123–128.

Collins, W. E., P. G. Contacos, and E. G. Guinn. 1967. Studies on the transmission of simian malarias. II: Transmission of the H strain of *Plasmodium knowlesi* by *Anopheles balababcensis balabacensis. Journal of Parasitology* 53(4): 841–844.

Collins, W. E., P. G. Contacos, E. G. Guinn, and J. R. Held. 1967. Studies on the transmission of simian malaria. III: Infection and transmission of *Plasmodium coatneyi* with *Anopoheles freeborni* and *A. balabacensis balababensis* mosquitoes. *Journal of Parasitology* 53(6): 1130–1134.

Collins, W. E., P. G. Contacos, E. G. Guinn, M. H. Jeter, and T. M. Sodeman. 1968. Monkey to man transmission of *Plasmodium falciparum* by *Anopheles freeborni* mosquitoes. *Journal of Parasitology* 54(6): 1166–1170.

Collins, W. E., P. G. Contacos, E. G. Guinn, and J, C. Skinner. 1973a. *Plasmodium simium* in the *Aotus trivirgatus* monkey. *Journal of Parasitology* 5(1): 49–51.

Collins, W. E., P. G. Contacos, and J. R. Jumper. 1972. Studies on the exoerythrocytic stages of simian malaria. VII: *Plasmodium simiovale. Journal of Parasitology* 58(1): 135–141.

———. 1974. Studies on the exoerythrocytic stages of simian malaria. IX: *Plasmodium simium. Journal of Parasitology* 60(4): 674–676.

Collins, W. E., P. G. Contacos, J. R. Jumper, C. S. Smith, and J. C. Skinner. 1973b. Studies on the exoerythrocytic stages of simian malaria. VIII: *Plasmodium knowlesi. Journal of Parasitology* 59(2): 344–352.

———. 1973d. Studies on human malaria in *Aotus* monkeys. III: Exoerythrocytic stages of the Salvador II strain of *Plasmodium vivax. Journal of Parasitology* 59(5): 859–866.

Collins, W. E., P. G. Contacos, W. A. Krotoski, and W. A. Howard. 1972. Transmission of four Central American strains of *Plasmodium vivax* from monkey to man. *Journal of Parasitology* 58(2): 332–335.

Collins, W. E., P. G. Contacos, J. C. Skinner, and E. G. Guinn. 1971a. Studies on the transmission of simian malaria. IV: Further studies on the transmission of *Plasmodium knowlesi* by *Anopheles balabacensis balabacensis* mosquitoes. *Journal of Parasitology* 57(5): 961–966.

______. 1967b. Fluorescent-antibody studies on simian malaria. I: Development of antibodies to *Plasmodium knowlesi. American Journal of Tropical Medicine and Hygiene* 16(1): 1–6.

Collins, W. E., P. G. Contacos, J. C. Skinner, A. Y. Huong, and W. Chin. 1982b. Studies on the Cambodian I strain of *Plasmodium falciparum* in *Aotus* monkeys. *Journal of Parasitology* 68(5): 877–883.

Collins, W. E., P. G. Contacos, P. S. Stanfill, and B. B. Richardson. 1974a. *Plasmodium hylobati* in the Rhesus monkey. *Journal of Parasitology* 60(6): 1038–1039.

Collins, W. E., P. G. Contacos, J. C. Skinner, P. S. Stanfill, and B. B. Richardson. 1981a. Susceptibility of Peruvian *Aotus* monkeys to infection with different species of *Plasmodium. American Journal of Tropical Medicine and Hygiene* 30(1): 26–30.

Collins, W. E., K. K. Grady, P. Millet, J. Sullivan, C. L. Morris, G. G. Galland, B. B. Richardson, and C. Yang. 1997a. Adaptation of a strain of *Plasmodium falciparum* from a Montagnard refugee to *Aotus* monkeys. *Journal of Parasitology* 83(6): 1174–1177.

Collins, W. E., G. M. Jeffery, and J. C. Skinner. 1964a. Fluorescent antibody studies in human malaria. I: Development of antibodies to *Plasmodium malariae. American Journal of Tropical Medicine and Hygiene* 13(1): 1–5.

Collins, W. E., J. R. Jumper, C. S. Smith, and J. C. Skinner. 1964b. Fluorescent antibody studies in human malaria. II: Development and persistence of antibodies to human malaria. *American Journal of Tropical Medicine and Hygiene* 13(2): 256–260.

———. 1964c. Fluorescent antibody studies in human malaria. III: Development of antibodies to *Plasmodium falciparum* in semi-immune patients. *American Journal of Tropical Medicine and Hygiene* 13(6): 777–782.

———. 1977. The exoerythrocytic states of *Plasmodium fragile* in *Macaca mulatta* monkeys. *Journal of Parasitology* 63(2): 226–231.

Collins, W. E., H. M. McClure, E. Strobert, V. Filipski, J. C. Skinner, P. S. Stanfill, B. B. Richardson, and C. Morris. 1990a. Infection of chimpanzees with the Uganda/ICDC strain of *Plasmodium malariae. American Journal of Tropical Medicine and Hygiene* 42(2): 99–103.

Collins, W. E., L. H. Miller, R. H. Glew, P. G. Contacos, W. A. Howard, and D. J. Wyler. 1973c. Transmission of three strains of *Plasmodium falciparum* from monkey to man. *Journal of Parasitology* 59(5): 855–858.

Collins, W. E., C. L. Morris, B. B. Richardson, J. Sullivan, and G. G. Galland. 1994. Further studies on the sporozoite transmission of the Salvador I strain of *Plasmodium vivax*. *Journal of Parasitology* 80(4): 512–517.

Collins, W. E., F. A. Neva, E. Chaves-Carballo, P. S. Stanfill, and B. B. Richardson. 1973d. Studies on human malaria in *Aotus* monkeys. II: Establishment of a strain of *Plasmodium falciparum* from Panama. *Journal of Parasitology* 59(4): 609–612.

Collins, W. E., M. Pappaioanou, H. M. McClure, R. B. Swenson, E. Strobert, J. C. Skinner, V. Filipski, P. S. Stanfill, F. H. Collins, and C. C. Campbell. 1987a. Infection of chimpanzees with Nigerian I/CDC strain of *Plasmodium ovale*. *American Journal of Tropical Medicine and Hygiene* 37(3): 455–459.

Collins, W. E., B. B. Richardson, J. Sullivan, C. L. Morris, and G. G. Galland. 1997b. Infection of *Anopheles freeborni* mosquitoes on New World monkeys infected with the Ugandan I/CDC strain of *Plasmodium malariae*. *Journal of Parasitology* 83(6): 1099–1103.

Collins, W. E., J. Sattabongkot, R. A. Wirtz, J. C. Skinner, J. R. Broderson, P. G. Millet, C. L. Morris, B. B. Richardson, J. Sullivan, and V. K. Filipski. 1992a. Development of a polymorphic strain of *Plasmodium vivax* in monkeys. *Journal of Parasitology* 78(3): 485–491.

Collins, W. E., I. K. Schwartz, J. C. Skinner, and J. R. Broderson. 1984. Studies on the Ugandan I/CDC strain of *Plasmodium malariae* in Bolivian *Aotus* monkeys. *Journal of Parasitology* 70(5): 677–681.

Collins, W. E., J. C. Skinner, J. R. Broderson, A. Y. Huong, P. C. Mehaffey, P. S. Stanfill, and B. B. Sutton. 1986. Infections of *Aotus azarae boliviensis* monkeys with different strains of *Plasmodium falciparum*. *Journal of Parasitology* 72(4): 525–530.

Collins, W. E., J. C. Skinner, J. R. Broderson, V. K. Filipski, C. M. Morris, P. S. Stanfill, and M. Warren. 1992b. Susceptibility of *Macaca fascicularis* monkeys from Mauritius to different species of *Plasmodium*. *Journal of Parasitology* 78(3): 505–511.

Collins, W. E., J. C. Skinner, J. R. Broderson, M. Pappaioanou, V. Filipski, B. B. Sutton, P. S. Stanfill, A. Y. Huong, J. Roberts, and C. Wilson. 1989b. The Uganda I/CDC strain of *Plasmodium malariae* in *Aotus lemurinus griseimembra* monkeys. *Journal of Parasitology* 75(1): 61–65.

Collins, W. E., J. C. Skinner, J. R. Broderson, B. B. Richardson, and P. S. Stanfill. 1989a. The Ugandan I/CDC strain of *Plasmodium malariae* in *Saimiri sciureus boliviensis*. *Journal of Parasitology* 75(2): 310–313.

Collins, W. E., J. C. Skinner, J. R. Broderson, B. B. Richardson, N. Shi-Fong Ma, and P. S. Stanfill. 1991. Infection of *Aotus vociferans* with different strains of *Plasmodium falciparum*. *Journal of Parasitology* 77(4): 562–567.

Collins, W. E., J. C. Skinner, P. G. Contacos, E. Guinn, and B. B. Richardson. 1973e. The influence of splenectomy on infections of *Plasmodium fieldi* in *Macaca mulatta* monkeys. *Journal of Parasitology* 59(1): 41–48.

Collins, W. E., J. C. Skinner, V. K. Filipski, J. R. Broderson, P. S. Stanfill, and C. L. Morris. 1990b. Transmission of *Plasmodium fragile* to *Saimiri* monkeys. *Journal of Parasitology* 76(5): 730–732.

Collins, W. E., J. C. Skinner, V. Filipski, C. Wilson, J. R. Broderson, and P. S. Stanfill. 1988a. Transmission of the OS strain of *Plasmodium inui* to *Saimiri sciureus boliviensis* and *Aotus azarae boliviensis* monkeys by *Anopheles dirus* mosquitoes. *Journal of Parasitology* 74(3): 502–503.

Collins, W. E., J. C. Skinner, A. Y. Huong, J. R. Broderson, B. B. Sutton, and P. Mehaffey. 1985a. Studies on a newly isolated strain of *Plasmodium brasilianum* in *Aotus* and *Saimiri* monkeys. *Journal of Parasitology* 71(6): 767–770.

Collins, W. E., J. C. Skinner, W. A. Krotoski, F. B. Cogswell, R. W. Gwadz, J. R. Broderson, N. S. F. Ma, P. Mehaffy, and B. B. Sutton. 1985b. Studies on the north Korean strain of *Plasmodium vivax* in *Aotus* monkeys and different anophelines. *Journal of Parasitology* 71(1): 20–27.

Collins, W. E., J. C. Skinner, and P. Mehaffey. 1985. Infection of *Aotus azarae boliviensis* monkeys with the RO strain of *Plasmodium cynomolgi*. *Journal of Parasitology* 71(6): 848–849.

Collins, W. E., P. S. Stanfill, B. B. Richardson, and C. S. Smith. 1974b. Transmission of *Plasmodium fragile* to *Aotus trivirgatus* monkeys. *Journal of Parasitology* 60(4): 719–720.

Collins, W. E., J. C. Skinner, M. Pappaioanou, J. R. Broderson, V. K. Filipksi, H. M. McClure, E. Strobert, B. B. Sutton, P. S. Stanfill, and A. Y. Huong. 1988b. Sporozoite-induced infections of the Salvador I strain of *Plasmodium vivax* in *Saimiri sciureus boliviensis* monkeys. *Journal of Parasitology* 74(4): 582–585.

Collins, William E., Jimmie C. Skinner, Marguerite Pappaioanou, J. Roger Broderson, Harold M. McClure, Elizabeth Strobert, Bettye B. Sutton, Peggy S. Stanfill, Valerie Filipski, and Carlos C. Campbell. 1987b. Chesson strain *Plasmodium vivax* in *Saimiri sciureus boliviensis* monkeys. *Journal of Parasitology* 73(5): 929–934.

Collins, William E., Jimmie C. Skinner, Marguerite Pappaioanou, J. Roger Broderson, and Philip Mehaffe. 1986. The sporogonic cycle of *Plasmodium reichenowi*. *Journal of Parasitology* 72(2): 292–298.

Collins, W. E., J. C. Skinner, M. Pappaioanou, J. R. Broderson, N. Shi-Fong Ma, V. Filipski, P. S. Stanfill, and L. Rogers. 1988c. Infection of Peruvian *Aotus nancymai* monkeys with different strains of *Plasmodium falciparum*, *P. vivax*, and *P. malariae*. *Journal of Parasitology* 74(3): 392–398.

Collins, W. E., J. C. Skinner, M. Pappaionou, J. Roger Broderson, N. Shi-Fong Ma, P. S. Stanfill, and V. Filipski. 1987c. Transmission of *Plasmodium simium* to *Aotus nancymai*, *A. vociferans*, *A. azarae boliviensis*, and *Saimiri sciureus boliviensis* monkeys. *Journal of Parasitology* 73(3): 653–655.

Collins, W. E., J. C. Skinner, M. Pappaionou, N. Shi-Fong Ma, J. R. Broderson, B. B. Sutton, and P. S. Stanfill. 1987d. Infection of *Aotus vociferans* (Karyotype V) monkeys with different strains of *Plasmodium vivax*. *Journal of Parasitology* 7(3): 536–540.

Collins, W. E., J. C. Skinner, B. B. Richardon, and A. Y. Huong. 1981b. Development of *Plasmodium inui* in the *Aotus trivirgatus* monkey. *Journal of Parasitology* 67(2): 222–225.

Collins, W. E., J. C. Skinner, B. B. Richardson, and P. S. Stanfill. 1975a. Transmission of *Plasmodium cynomolgi* to *Aotus trivirgatus* monkeys. *Journal of Parasitology* 61(1): 146–148.

———. 1975b. Studies on the transmission of simian malaria VI. Mosquito infection and sporozoite transmission of *Plasmodium fragile*. *Journal of Parasitology* 61(4): 718–721.

Collins, W. E., J. C. Skinner, P. S. Stanfill, and B. B. Richardson. 1978. Studies on the Burma (Thau.) strain of *Plasmodium falciparum* in *Aotus trivirgatus* monkeys. *Journal of Parasitology* 64(3): 497–500.

Collins, W. E., P. S. Stanfill, B. B. Richardson, and C. S. Smith. 1974b. Transmission of *Plasmodium fragile* to *Aotus trivirgatus* monkeys. *Journal of Parasitology* 60(4): 719–720.

Collins, W. E., J. Sullivan, G. G. Galland, J. W. Barnwell, D. Nace, A. Williams, T. Williams, and A. Bounngaseng. 2004a. Rio Meta strain of *Plasmodium vivax* in New World monkeys and anopheline mosquitoes. *Journal of Parasitology* 90(4): 685–688.

Collins, W. E., J. S. Sullivan, D. Nace, T. Williams, A. Williams, G. G. Galland, and J. W. Barnwell. 2004b. Additional observations on the sporozoite transmission of *Plasmodium knowlesi* to monkeys. *Journal of Parasitology* 90(4): 866–867.

Collins, W. E., M. Warren, P. G. Contacos, J. C. Skinner, and B. B. Richardson. 1979b. Infectivity of *Plasmodium simium* to *Aotus trivirgatus* monkeys and different anophelines. *Journal of Parasitology* 65(6): 870–874.

Collins, W. E., M. Warren, P. G. Contacos, J. C. Skinner, B. B. Richardson, and T. S. Kearse. 1980a. The Chesson strain of *Plasmodium vivax* in *Aotus* monkeys and anopheline mosquitoes. *Journal of Parasitology* 66(3): 488–497.

Collins, W. E., M. Warren, and G. G. Galland. 1999. Studies on infections with the Berok strain of *Plasmodium cynomolgi* in monkeys and mosquitoes. *Journal of Parasitology* 85(2): 268–272.

Collins, W. E., M. Warren, J. C. Skinner, W. Chin, and B. B. Richardson. 1977. Studies on the Santa Lucia (El Salvador) strain of *Plasmodium falciparum* in *Aotus trivirgatus* monkeys. *Journal of Parasitology* 63(1): 52–56.

Collins, W. E., M. Warren, J. C. Skinner, A. Y. Huong, and P. Nguyen-Dinh. 1983b. Observations on the infectivity of two strains of *Plasmodium vivax* from Vietnamese refugees to *Aotus* monkeys and anopheline mosquitoes. *Journal of Parasitology* 69(4): 689–695.

Collins, W. E., M. Warren, J. C. Skinner, and B. B. Richardson. 1980b. Studies on the west Pakistan strain of *Plasmodium vivax* in *Aotus* monkeys and anopheline mosquitoes. *Journal of Parasitology* 66(5): 780–785.

Collins, W. E., M. Warren, J. C. Skinner, B. B. Richardson, and W. Chin. 1979c. Studies on the west African I strain of *Plasmodium falciparum* in *Aotus trivirgatus* monkeys. *Journal of Parasitology* 65(5): 763–767.

Collins, W. E., M. Warren, J. C. Skinner, B. B. Richardson, and T. S. Kearse. 1979d. Effect of sequential infection with *Plasmodium vivax* and *P. falciparum* in the *Aotus trivirgatus* monkey. *Journal of Parasitology* 65(4): 605–608.

Collins, W. E., M. Warren, J. S. Sullivan, and G. G. Galland. 2001. *Plasmodium coatneyi*: Observations on periodicity, mosquito infection, and transmission to *Macaca mulatta* monkeys. *American Journal of Tropical Medicine and Hygiene* 64(3): 101–110.

Collins, W. E., M. Warren, J. S. Sullivan, G. G. Galland, D. Nace, A. Williams, T. Williams, and J. W. Barnwell. 2005. Studies on two strains of *Plasmodium cynomolgi* in New World monkeys and mosquitoes. *Journal of Parasitology* 91(2): 280–283.

Collins, W. E., M. Warren, J. S. Sullivan, G. G. Galland, E. Strobert, D. Nace, A. Williams, T. Williams, and J. W. Barnwell. 2006. Studies on sporozoite-induced and chronic infections with *Plasmodium fragile* in *Macaca mulatta* and New World monkeys. *Journal of Parasitology* 92(5): 1019–1026.

Coluzzi, M., V. Petrarca, and M. A. di Deco. 1985. Chromosomal inversion and intergradation and incipient speciation. *Anopheles gambiae. Bolletino di Zoologia* 52(1/2): 45–63.

Coluzzi, M., A. Sabatini, V. Petrarca, and M. A. di Deco. 1979. Chromosomal differentiation and adaptation to human environments in the *Anopheoes gambaiae* complex. *Transactions of the Royal Society of Tropical Medicine and Hygiene* 73(5): 483–497.

Contacos, P. G. 1970. Primate malarias: Man and monkeys. *Journal of Wildlife Diseases* 6: 323–328.

Contacos, P. G., and G. R. Coatney. 1963. Experimental adaptation of simian malarias to abnormal hosts. *Journal of Parasitology* 49(6): 912–918.

Contacos, P. G., G. R. Coatney, J. S. Lunn, and W. Chin. 1965. Resistance to cycloguanil pamoate (CI-501) by *falciparum* malaria in West Pakistan. *American Journal of Tropical Medicine and Hygiene* 14(6): 925–926.

Contacos, P. G., G. R. Coatney, J. S. Lunn, and J. W. Kilpatrick. 1964. The antimalarial activity of CI-501 (Camolar®) against *falciparum* malaria. *American Journal of Tropical Medicine and Hygiene* 13(3): 386–390.

Contacos, P. G., G. R. Coatney, T. C. Orihel, W. E. Collins, W. Chin, and M. H. Jeter. 1970. Transmission of *Plasmodium schwetzi* from the chimpanzee to man by mosquito bite. *American Journal of Tropical Medicine and Hygiene* 19(2): 190–195.

Contacos, P. G., and W. E. Collins. 1968. *Falciparum* malaria transmissible from monkey to man by mosquito bite. *Science* 161(3836): 56.

Contacos, P. G., W. E. Collins, G. M. Jeffery, W. A. Krotoski, and W. A. Howard. 1972. Studies on the characterization of *Plasmodium vivax* strains from Central America. *American Journal of Tropical Medicine and Hygiene* 21(5): 707–712.

Contacos, P. G., H. A. Elder, and G. R. Coatney. 1963. Therapeutic trials of a dihydroxynaphthalene (377C54) against *Plasmodium falciparum* and *Plasmodium vivax* infections in human volunteers. *American Journal of Tropical Medicine and Hygiene* 12(4): 513–518.

Contacos, P. G., H. A. Elder, G. R. Coatney, and C. Genther. 1962. Man to man transfer of two strains of *Plasmodium cynomolgi* by mosquito bite. *American Journal of Tropical Medicine and Hygiene* 11(2): 186–193.

Contacos, P. G., J. S. Lunn, G. R. Coatney, J. W. Kilpatrick, and F. W. Jones. 1963. Quartan-type malaria parasite of New World monkeys transmissible to man. *Science* 142(3593): 676.

Coluzzi, M., V. Petrarca, and M. A. di Deco. 1985. Chromosomal inversion and intergradation and incipient speciation in *Anopheles gambiae. Bolletino di zoologia* 52(1/2): 45–63.

Coluzzi, M., A. Sabatini, V. Petrarca, and M. A. di Deco. 1979. Chromosomal differentiation and adaptation to human environments in the *Anopheles gambiae* complex. *Transactions of the Royal Society of Tropical Medicine and Hygiene* 73(5): 483–497.

Conn, J. E., R. C. Wilkerson, M. N. O. Segura, R. T. L. de Souza, C. D. Schlichting, R. A. Wirtz, and M. M. Póvoa. 2002. Emergence of a new Neotropical malaria vector facilitated by human migration and changes in land use. *American Journal of Tropical Medicine and Hygiene* 66(1): 18–22.

Conrad, R. 1975. The planter class and the debate over Chinese immigration to Brazil, 1850–1893. *International Migration Review* 9: 41–55.

Cook, N. D. 1998. *Born to die: Disease and New World conquest, 1492–1650*. Cambridge: Cambridge University Press.

Cooper, W. C., A. V. Myatt, T. Hernandez, G. M. Jeffrey, and G. R. Coatney. 1953. Studies in human malaria. XXXI: Comparison of primaquine, isopentaquine, SN-3883, and pamaquine as curative agents against Chesson strain of *vivax* malaria. *American Journal of Tropical Medicine and Hygiene* 2(6): 949–957.

Corbett, J. B. 2001. Women, scientists, and agitators: Magazine portrayals of Rachel Carson and Theo Colborn. *Journal of Communication* 51(4): 720–749.

Cormier, L. A. 2010. The historical ecology of human and wild primate malarias in the New World. *Diversity* 2(2): 256–280, *www.mdpi.com/1424-2818/2/2/256/* (accessed August 5, 2010).

———. 2003. *Kinship with monkeys: The Guajá foragers of Eastern Amazonia*. Historical Ecology Series. New York: Columbia University Press.

———. 2006. A preliminary review of Neotropical primates in the subsistence and symbolism of indigenous lowland South American peoples. *Ecological and Environmental Anthropology* 2: 14–32.

Cornejo, O. E., and A. A. Escalante. 2006. The origin and age of *Plasmodium vivax*. *Trends in Parasitology* 22(12): 558–563.

Corredor, V., and V. Enea. 1993. Plasmodial ribosomal RNA as phylogenetic probe: A cautionary note. *Molecular Biology and Evolution* 10(4): 924.

Coura, J. R., M. Suárez-Mutis, and S. Ladeia-Andrade. 2006. A new challenge for malaria control in Brazil: Asymptomatic *Plasmodium* infection—a review. *Memoria Instituto Oswaldo Cruz* 101: 229–237.

Cox-Singh, J., T. M. E. Davis, K.-S. Lee, S. S. G. Shamsul, A. Matusop, S. Ratnam, H. A. Rahman, D. J. Conway, and B. Singh. 2008. *Plasmodium knowlesi* malaria in humans is widely distributed and potentially life threatening. *Clinical Infectious Diseases* 46: 165–171.

Craige, B., Jr., R. Jones, Jr., C. M. Whorton, T. N. Pullman, A. S. Alving, and L. Eichelberger. 1947. Clinical standardization of pamaquin (plasmochin) in mosquito-induced *vivax* malaria, Chesson strain: A preliminary report. *American Journal of Tropical Medicine and Hygiene* s1-2(3): 309–315.

Crosby, A. W. 1994. *Germs, seeds, and animals: Studies in ecological history*. Armonk: M. E. Sharpe.

Cserti, C. M., and W. H. Dzik. 2007. The ABO blood group and *Plasmodium falciparum* malaria. *Blood* 110(7): 2250–2258, *http://bloodjournal.hematologylibrary.org/cgi/reprint/110/7/2250.pdf* (accessed December 16, 2009).

Currat, M., and L. Excoffier. 2004. Modern humans did not admix with Neanderthals during their range expansion into Europe. *PLoS Biology* 2(12): e421, *www.plosbiology.org/article/info%3Adoi%2F10.1371%2Fjournal.pbio.0020421* (accessed February 12, 2011).

Da Fonseca, F. 1951. Plasmódio de primata do Brasil. *Memórias do Instituto Oswaldo Cruz* 49: 543–551, *http://memorias.ioc.fiocruz.br/pdf/Tomo49/tomo49(fu)_543-551.pdf* (accessed November 25, 2008).

Davies, C. R., J. M. Ayres, C. Dye, and L. M. Deane. 1991. Malaria infection rate of Amazonian primates increases with body weight and group size. *Functional Ecology* 5(5): 655–662.

Davison, B. B., F. B. Cogswell, G. B. Baskin, K. P. Falkenstein, E. W. Henson, and D. J. Krogstad. 2000. Placental changes associated with fetal outcome in *Plasmodium coatneyi*/Rhesus monkey model of malaria in pregnancy. *American Journal of Tropical Medicine and Hygiene* 63(3): 158–173.

Davison, B. B., F. B. Cogswell, G. B. Baskin, K. P. Falkenstein, E. W. Henson, A. F. Tarantal, and D. J. Krogstad. 1998. *Plasmodium coatneyi* in the Rhesus monkey (*Macaca mulatta*) as a model of malaria in pregnancy. *American Journal of Tropical Medicine and Hygiene* 59(2): 189–201.

Dawkins, R. 1999. *The extended phenotype*. Oxford: Oxford University Press.

de Jong, R., and B. G. J. Knols. 1995. Olfactory responses of host-seeking *Anopheles gambiae s.s.* Giles (Diptera: Culicidae). *Acta Tropica* 59(4): 333–335.

de Thoisy, B., I. Vogel, J.-M. Reynes, J. F. Pouliquen, B. Carme, M. Kazanji, and J.-C. Vie. 2001. Health evaluation of translocated free-ranging primates. *American Journal of Primatology* 5(1): 1–16.

de Vos, J., and P. Sondaar. 1994. Dating hominid sites in Indonesia. *Science* 266(5191): 1726–1727.

Deane, L. M. 1964. Studies on simian malaria in Brazil. *Bulletin of the World Health Organization* 31(5): 752–753.

———. 1992. Simian malaria in Brazil. *Memoria Instituto Oswaldo Cruz* 87: 1–20.

Deane, L. M., J. A. Feirreira Neto, M. Paumgartten Deane, and I. P. S. Silveira. 1970. *Anopheles (Kerteszia) cruzii* a natural vector of malaria parasites *Plasmodium simium* and *Plasmodium brasilianum*. *Transactions of the Royal Society of Tropical Medicine and Hygiene* 64: 647.

Deane, L. M., J. Feirreira Neto, and I. P. S. Silveira. 1966. Experimental infection of splenectomized squirrel monkey, *Saimiri sciureus* with *Plasmodium vivax*. *Transactions of the Royal Society of Tropical Medicine and Hygiene* 60(6): 811–812.

Deane, L. M., M. Paumgartten Deane, and J. Ferreira Neto. 1966. Studies on transmission of simian malaria and on a natural infection of man with *Plasmodium simium* in Brazil. *Bulletin of the World Health Organization* 35(5): 805–808.

De Arruda, M., E. H. Nardin, R. S. Nussenzweig, and A. H. Cochrane. 1989. Sero-epidemiological studies of malaria in Indian tribes and monkeys of the Amazon Basin of Brazil. *American Journal of Tropical Medicine and Hygiene* 41: 379–385.

DeGowin, R. L., and R. D. Powell. 1965. Drug resistance of a strain of *Plasmodium falciparum* from Malaya. *American Journal of Tropical Medicine and Hygiene* 14(4): 519–528.

Denevan, W. M. 1992a. The pristine myth: The landscape of the Americas in 1492. *Annals of the Association of American Geographers* 82: 369–385.

———. 1992b. *The native population of the Americas in 1492*. Madison: University of Wisconsin Press.

Dern, R. J., E. Beutler, J. Arnold, A. Lorincz, M. Block, and A. S. Alving. 1955. Toxicity studies of pyrimethamine (daraprim). *American Journal of Tropical Medicine and Hygiene* 4(2): 217–220.

Desowitz, R. S. 1991. *The malaria capers: More tales of people, research, and reality*. New York: W. W. Norton and Company.

Dillehay, T. D., C. Ramírez, M. Pino, M. B. Collins, J. Rossen, and J. D. Pino-Navarro. 2008. Monte Verde: Seaweed, food, medicine, and the peopling of South America. *Science* 320(5877): 784–786.

Dissanaike, A. S. 1965. Simian malaria parasites of Ceylon. *Bulletin of the World Health Association* 32(4): 593–597.

Diuk-Wasser, M. A., M. B. Touré, G. Dolo, M. Bagayoko, N. Soboga, I. Sissoko, S. F. Traoré, and C. E. Taylor. 2007. Effect of rice cultivation patterns on malaria vector abundance in rice-growing villages in Mali. *American Journal of Tropical Medicine and Hygiene* 76(5): 869–874.

Dobyns, H. F. 1983. *Their number become thinned: Native American population dynamics in eastern North America.* Knoxville: University of Tennessee Press.

———. 1993. Disease transfer at contact. *Annual Review of Anthropology* 22: 273–291.

Donnelly, M. J., M. C. Licht, and T. Lehmann. 2001. Evidence for recent population expansion in the evolutionary history of the malaria vectors *Anopheles arabiensis* and *Anopheles gambiae*. *Molecular Biology and Evolution* 18(7): 1353–1364.

Donovan, J. C., W. S. Stokes, R. Montrey, and H. Rozmiarek. 1983. Hematologic characterization of naturally occurring malaria (*Plasmodium inui*) in the cynomolgus monkey (*Macaca fascicularis*). *Laboratory Animal Science* 33(1): 86–89.

Drummond Alexei J., Simon Y. W. Ho, Matthew J. Phillips, and Andrew Rambaut. 2006. Relaxed phylogenetics and dating with confidence. *PLoS Biology* 4(5): e88. *www.plosbiology.org/article/info:doi/10.1371/journal.pbio.0040088* (accessed August 2, 2010).

Duarte, A. M., Ribeiro de Castro, R. dos Santos Malafronte, C. Cerutti, Jr., I. Curado, B. R. de Paiva, A. Y. Maeda, T. Yamasaki, M. E. L. Summa, D. do Valle Dutra de Andrade Neves, S. Gomes de Oliveira, and A. de Castro Gomes. 2008. Natural *Plasmodium* infectious in Brazilian wild monkeys: Reservoirs for human infections? *Acta Tropica* 107(2): 179–185.

Dunn, F. L. 1965. On the antiquity of malaria in the western hemisphere. *Human Biology* 37: 38–43.

———. 1993. Malaria. In *The Cambridge World History of Human Disease*, K. F. Kiple, Ed., 855–862. Cambridge: Cambridge University Press.

Dunn, F. L., and F. L. Lambrecht. 1963. Hosts of *Plasmodium brasilianum* Gonder and von Berenberg-Glosser, 1908. *Journal of Parasitology* 49(2): 316–319.

Duval, L., M. Fourment, E. Nerrienet, D. Rousset, S. A. Sadeuh, S. M. Goodman, N. V. Andriaholinirina, M. Randrianarivelojosia, R. E. Paul, V. Robert, F. J. Ayala, and F. Ariey. 2010. African apes as reservoirs of *Plasmodium falciparum* and the origin and diversification of the *Laverania* subgenus. *Proceedings of the National Academy of Sciences, United States of America* 107(23): 10561–10566.

Duval, L., E. Nerrienet, D. Rousset, S. A. Sadeuh Mba, S. Houze, M. Fourment, J. Le Bras, V. Robert, and F. Ariey. 2009. Chimpanzee malaria parasites related to *Plasmodium ovale* in Africa. *Public Library of Science One* 4(5) (May 13), *www.pubmedcentral.nih.gov/articlerender.fcgi?artid=2677663* (accessed September 3, 2009).

Eberle, R., and J. Hilliard. 1995. The simian herpesviruses. *Infectious Agents and Disease* 4: 55–70.

Edwards, J. G. 2004. DDT: A case study in scientific fraud. *Journal of American Physicians and Surgeons* 9(3): 83–88.

Emmons, L. H., and F. Feer. 1997. *Neotropical rainforest mammals: A field guide.* Chicago: University of Chicago Press.

Engwerda, C. R., L. Beattie, and F. H. Amante. 2005. The importance of the spleen in malaria. *Trends in Parasitology* 21(2):75–80.

Eronen, J. T., and L. Rook. 2004. The Mio-Pliocene European primate fossil record: Dynamics and habitat tracking. *Journal of Human Evolution* 47(5): 323–341.

Escalante, A. A., and F. J. Ayala. 1994. Phylogeny of the malarial genus *Plasmodia*, derived from rRNA gene sequences. *Proceedings of the National Academy of the Sciences, United States of America* 91: 11373–11377.

Escalante, A. A., E. Barrio, and F. J. Ayala. 1995. Evolutionary origin of human and primate malarias: Evidence from the circumsporozoite protein gene. *Molecular Biology and Evolution* 12(4): 616–626.

Escalante, A. A., O. E. Cornejo, D. E. Freeland, A. C. Poe, E. Durrego, W. E. Collins, and A. A. Lal. 2005. A monkey's tale: The origin of *Plasmodium vivax* as a human malaria parasite. *Proceedings of the American Academy of Sciences,United States of America* 102(6): 180–185.

Escalante, A. A., D. E. Freeland, W. E. Collins, and A. A. Lal. 1998. The evolution of primate malaria parasites based on the gene encoding cytochrome b from the linear mitochondrial genome. *Proceedings of the National Academy of Sciences, United States of America* 95(14): 8124–8129.

Escalante, A. A., I. F. Goldman, P. De Rijk, R. De Wachter, W. E. Collins, S. H. Qari, and A. A. Lal. 1997. Phylogenetic study of the genus *Plasmodium* based on the secondary structure-based alignment of the small subunit ribosomal RNA. *Molecular and Biochemical Parasitology* 90(1): 317–321.

Evans, P. D., N. Mekel-Bobrov, E. J. Vallender, R. R. Hudson, and B. T. Lahn. 2006. Evidence that the adaptive allele of the brain size gene *microcephalin* introgressed into *Homo sapiens* from an archaic *Homo* lineage. *Proceedings of the National Academy of Sciences, United States of America* 103(48): 18178–18183.

Ewald, P. W. 1983. Host-parasite relations, vectors, and the evolution of disease severity. *Annual Review of Ecology and Systematics* 14: 465–485.

———. 1994. *The evolution of infectious disease*. Oxford: Oxford University Press.

Eyles, D. E. 1963. The species of simian malaria: Taxonomy, morphology, life cycle, and geographical distribution of the monkey species. *Journal of Parasitology* 49(6): 866–887.

Eyles, D. E., G. R. Coatney, and M. E. Getz. 1960. *Vivax*-type malaria parasite of macaques transmissible to man. *Science* 131(3416): 1812–1813.

Eyles, D. E., F. L. Dunn, M. Warren, and E. Guinn. 1963. *Plasmodium coatneyi* from the Phillipines. *Journal of Parasitology* 49(6): 1038.

Eyles, D., Y. L. Fong, F. L. Dunn, E. Guinn, M. Warren, and A. A. Sandosham. 1964. *Plasmodium youngi* n. sp., a malaria parasite of the Malayan gibbon, *Hylobates lar. American Journal of Tropical Medicine and Hygiene* 13(2): 248–255.

Eyles, D. E., Y. L. Fong, M. Warren, E. Guinn, A. A. Sandosham, and R. H. Wharton. 1962. *Plasmodium coatneyi*, a new species of primate malaria from Malaya. *American Journal of Tropical Medicine and Hygiene* 11(5): 597–604.

Eyles, D. E., and M. Warren. 1962. *Plasmodium inui* in Sulawesi. *Journal of Parasitology* 48(5): 739.

Falk, D., C. Hildebolt, K. Smith, M. J. Morwood, T. Sutikna, P. Brown, J. E. Wayhu Saptomo, B. Brunsden, and F. Prior. 2005. The brain of LB1, *Homo floresiensis. Science* 308(5719): 242–245.

Fandeur, T., B. Volney, C. Peneau, and B. de Thoisy. 2000. Monkeys of the rainforest in French Guiana are natural reservoirs for *P. brasilianum/P. malariae* malaria. *Journal of Parasitology* 120(Pt 1): 11–21.

Farmer, P. 2001. *Infections and inequalities: The modern plagues*. Berkeley and Los Angeles: University of California Press.

Feldman, D. A. 1990. Assessing viral, parasitic, and sociocultural cofactors affecting HIV-1transmission in Rwanda. In *Culture and AIDS*, D. A. Feldman, Ed., 45–54. New York: Praeger.

Feng, X., J. M. Carlton, D. A. Joy, J. Mu, T. Furuya, B. B. Suh, Y. Wang, J. W. Barnwell, and X.-Z. Su. 2003. Single-nucleotide polymorphisms and genome diversity in *Plasmodium vivax*. *Proceedings of the National Academy of Sciences, United States of America* 100(14): 8502–8507.

Ferguson, H. M., and A. F. Read. 2002. Genetic and environmental determinants of malaria parasite virulence in mosquitoes. *Proceedings of the Royal Society of London: Biological Sciences* 269(1497): 1217–1224.

Fiennes, R. 1967. *Zoonoses of primates, the epidemiology and ecology of simian diseases in relation to man*. Ithaca, NY: Cornell University Press.

Fischer, P. R., and P. Boone. 1998. Short report: Severe malaria associated with blood group. *American Journal of Tropical Medicine and Hygiene* 58(1): 122–123.

Fleagle, J. G. 1999. *Primate adaptation and evolution*, second edition. New York: Academic Press.

Fleagle, J. G., and E. L. Simons. 1982. The humerus of *Aegpytopithecus zeuxis*: A primitive anthropoid. *American Journal of Physical Anthropology* 59(2): 175–193.

Fooden, J. 1963. A revision of the woolly monkeys (genus Lagothrix). *Journal of Mammology* 44(2): 213–247.

———. 1994. Malaria in macaques. *International Journal of Primatology* 15(4): 573–592.

Fowkes, F. J. I., S. J. Allen, A. Allen, M. P. Alpers, D. J. Weatherall, and K. P. Day. 2008. Increased microerythrocyte count in homozygous α (+)– thalassaemia contributes to protection against severe malarial anaemia. *PLoS Medicine* 5(3), *http://medicine.plosjournals.org/archive/1549-1676/5/3/pdf/10.1371_journal.pmed.0050056-L.pdf* (accessed May 31, 2008).

Frankenburg, F. R., and R. J. Balessarini. 2008. Neurosyphilis, malaria, and the discovery of antipsychotic agents. *Harvard Review of Psychiatry* 16(5): 299–307.

Fremount, H. N., and L. H. Miller. 1975. Deep vascular schizogony in *Plasmodium fragile*: Organ distribution and ultrastructure of erthrocytes adherent to vascular endothelium. *American Journal of Tropical Medicine and Hygiene* 24(1): 1–8.

Fremount, H. N., and R. N. Rossan. 1990. Anatomical distribution of developing trophozoites and schizonts of *Plasmodium vivax* in *Aotus lemurinus lemurinus* and *Saimiri sciureus*. *Journal of Parasitology* 76(3): 428–430.

Fretwell, S. D., and H. L. Lucas, Jr. 1969. On territorial behavior and other factors influencing habitat distribution in birds. I: Theoretical development. *Acta Biotheoretica* 19(1): 16–36.

Fürstenwerth, H. 2003. On morality and chemistry. *Poiesis & Praxis: International Journal of Ethnics of Science and Technology Assessment* 2: 53–71.

Galinski, M. R., and J. W. Barnwell. 2009. Monkey malaria kills four humans. *Trends in Parasitology* 2(5): 200–209.

Gao, F., E. Bailes, D. L. Robertson, Y. Chen, C. M. Rodenburg, S. F. Michael, L. B. Cummins, L. O. Arthur, M. Peeters, G. M. Shaw, P. M. Sharp, and B. H. Hahn. 1999. Origin of HIV-1 in the chimpanzee *Pan troglodytes troglodytes*. *Nature* 397: 436–441.

Garamszegi, L. Z. 2009. Patterns of co-speciation and host switching in primate malaria parasites. *Malaria Journal* 8(May 22), *www.pubmedcentral.nih.gov/articlerender.fcgi?artid=2689253* (accessed September 3, 2009).

Garnham, P. C. C. 1963. Distribution of simian malaria parasites in various hosts. *Journal of Parasitology* 49(6): 905–911.

Garnham, P. C. C., and K. L. Kuttler. 1980. A malaria parasite of the white-tailed deer (*Odocoileus virginianus*) and its relation to known species of *Plasmodium* in other ungulates. *Proceedings of the Royal Society of London, Series B, Biological Sciences* 206: 395–402.

Garnham, P. C. C., V. Molinari, and P. G. Shute. 1962. Differential diagnosis of *basianellii* and *vivax* malaria. *Bulletin of the World Health Organization* 27(2): 199–202.

Geiman, Q., and W. Siddiqui. 1969. Susceptibility of a New World monkey to *Plasmodium malariae* from man. *American Journal of Tropical Medicine and Hygiene* 18(3): 351–354.

Genton, B., F. Ai-Yaman, C. S. Mgone, N. Alexander, M. Paniu, M. P. Alpers, and D. Mokela. 1995. Ovalocytosis and cerebral malaria. *Nature* 378: 564–565.

Gilks, C. 2001. Man, monkeys, and malaria. *Philosophical Transactions of the Royal Society of London B* 356: 921–922.

Goebel, T., M. R. Waters, and D. H. O'Rourke. 2008. The late Pleistocene dispersal of modern humans in the Americas. *Science* 319, 1497–1502.

Good, K. R. 1987. Limiting factors in Amazonian ecology. In *Food and evolution: Toward a theory of human food habits*, Marvin Harris and Eric B. Ross, Eds., 407–421. Philadelphia: Temple University Press.

Gordon, A. D., L. Nevell, and B. Wood. 2008. The *Homo floresiensis* cranium (LB1): Size, scaling, and early *Homo affinities. Proceedings of the National Academy of Sciences, United States of America* 105(12): 4650–4655.

Gould, D. J., and F. C. Cadigan. 1966. *Falciparum* malaria: Transmission to the gibbon by *Anopheles balabacensis. Science* 153(3742): 1384.

Green, K. M. 1976. The nonhuman primate trade in Colombia. In *Neotropical primates: Field studies and conservation*, Eds., R. W. Thorington and P. B. Heltne, 85–95. Washington D.C.: National Academy of Sciences.

Green, R. C. 2005. Sweet potato transfers in Polynesian prehistory. In *The sweet potato in Oceania: A reappraisal*, C. Ballard, P. Brown, R. M. Bourke, and T. Harwood. Eds., 43–62. Sydney, Australia: Oceania Publications.

Green, R. E., J. Krause, A. W. Briggs, T. Maricic, U. Stenzel, M. Kircher, N. Patterson, H. Li, W. Zhai, M. Hsi-Yang Fritz, N. F. Hansen, E. Y. Durand, A.-S. Malaspinas, J. D. Jensen, T. Marques-Bonet, C. Alkan, K. Prüfer, M. Meyer, H. A. Burbano, J. M. Good, R. Schultz, A. Aximu-Petri, A. Butthof, B. Höber, B. Höffner, M. Siegemund, A. Weihmann, C. Nusbaum, E. S. Lander, C. Russ, N. Novod, J. Affourtit, M. Egholm, C. Verna, P. Rudan, D. Brajkovic, Z. Kucan, I. Gusic, V. B. Doronichev, L. V. Golovanova, C. Lalueza-Fox, M. de la Rasilla, J. Fortea, A. Rosas, R. W. Schmitz, P. L. F. Johnson, E. E. Eichler, D. Falush, E. Birney, J. C. Mullikin, M. Slatkin, R. Nielsen, J. Kelso, M. Lachmann, D. Reich, and S. Pääbo. 2010. A draft sequence of the Neanderthal Genome. *Science* 328(5979): 710–722.

Gross, D. 1975. Protein capture and cultural development in the Amazon Basin. *American Anthropologist* 77: 526–549.

Gubler, D. J. 2002. The global emergence/resurgence of arboviral diseases as public health problems. *Archives of Medical Research* 33: 330–342.

Gysin, J. 1998. Animal models. In *Malaria: Parasites, pathogenesis, and protection.* Irwin W. Sherman, Ed., 419–443. Washington, D. C.: American Society for Microbiology Press.

Gysin, J., M. Hommel, and L. Pereira da Silva. 1980. Experimental infection of the squirrel monkey (*Saimiri sciureus*) with *Plasmodium falciparum*. *Journal of Parasitology* 66(6): 1003–1009.

Haggett, P. 1994. Geographical aspects of the emergence of infectious disease. *Geografiska Annaler, Series B, Human Geography* 76: 91–104.

Hagner, S. C., B. Misof, W. A. Maier, and H. Kampen. 2007. Bayesian analysis of new and old malaria parasite DNA sequence data demonstrates the need for more phylogenetic signal to clarify the descent of *Plasmodium falciparum*. *Parasitology Research* 101(3): 493–503.

Haldane, J. B. S. 1949. The rate of mutation of human genes. *Proceedings of the VIII International Congress of Genetics Hereditas* 35: 267–273.

Hamblin, M. T., and A. Di Rienzo. 2000. Detection of the signature of natural selection in humans: Evidence from the Duffy blood group locus. *American Journal of Human Genetics* 66(5): 1669–1679.

Harlow, H. F., R. O. Dodsworth, and M. K. Harlow. 1965. Total social isolation in monkeys. *Proceedings of the National Academy of Science, United States of America* 5(1): 90–97.

Harms, D. 1979. *The Belmont Report: Ethical principles and guidelines for research involving human subjects.* The National Commission for the Protection of Biomedical and Behavioral Research. Washington, D.C.: United States Department of Health, Education, and Welfare.

Harris, A. F., S. Rajatileka, and H. Ranson. 2010. Pyrethroid resistance in *Aedes aegypti* from Grand Cayman. *American Journal of Tropical Medicine and Hygiene* 83(2): 277–284.

Harris, M. 1974. *Cows, pigs, wars and witches: The riddles of culture.* New York: Random House.

Hayakawa, T., N. Arisue, T. Udono, H. Hirai, J. Sattabongkot, T. Toyama, T. Tsuboi, T. Horii, K. Tanabe. 2009. Identification of *Plasmodium malariae*, a human malaria parasite, in imported chimpanzees. *PLoS ONE* 4(10): e7412, *www.plosone.org/article/info%3Adoi%2F10.1371%2Fjournal.pone.0007412* (accessed August 5, 2010).

Haynes, G. 2002. *The early settlement of North America: The Clovis era.* New York: Cambridge University Press.

Heckenberger, M. J., A. Kuikuro, U. T. Kuikuro, J. C. Russell, M. Schmidt, C. Fausto, B. Francetto. 2003. Amazonia 1492: Pristine forest or cultural parkland? *Science* 301(5640): 1710–1714.

Hedges, S. B., and S. Kumar. 2003. Genomic clocks and evolutionary timescales. *Trends in Genetics* 19(4): 200–206.

Heizmann, E. P. J., and D. R. Begun. 2001. The oldest Eurasian hominoid. *Journal of Human Evolution* 41(5): 463–481.

Held, J. R., and P. G. Contacos. 1967. Studies of the exoerythrocytic stages of simian malaria II: *Plasmodium coatneyi*. *Journal of Parasitology* 53(5): 910–918.

Held, J. R., P. G. Contacos, and G. R. Coatney. 1967. Studies of the exoerythrocytic stages of simian malaria I: *Plasmodium fieldi*. *Journal of Parasitology* 53(2): 225–232.

Hernandez, T., A. V. Myatt, G. R. Coatney, and G. M. Jeffery. 1953. Studies in human malaria: Acquired resistance to pyrimethamine (daraprim) by the Chesson strain of *Plasmodium vivax*. *American Society of Tropical Medicine and Hygiene* 2(5): 797–803.

Hernández-Comancho, J., and R. W. Cooper. 1976. The nonhuman primates of Colombia. In *Neotropical primates: Field studies and conservation*, Eds., R. W. Thorington and P. B. Heltne, 35–69. Washington D.C.: National Academy of Sciences.

Herrel, N., F. P. Amerasinghe, J. Ensink, M. Mukhtar, W. van der Hoek, and F. Konradsen. 2004. Adult anopheline ecology and malaria transmission in irrigated areas of South Punjab, Pakistan. *Medical and Veterinary Entomology* 18(2): 141–152.

Hewitt, G. 2000. The genetic legacy of the Quaternary ice ages. *Nature* 405: 907–913.

Heyerdahl, T. 1950. *Kon Tiki: Across the Pacific by raft*. New York: Rand McNally.

Himeidan, Y. E., M. I. Elbashir, and I. Adam. 2004. Attractiveness of pregnant women to the malaria vector, *Anopheles arabiensis*, in Sudan. *Annals of Tropical Medicine and Parasitology* 98(6): 631–633.

Hodgkinson, R., K. O. Courtney, and M. Haggerty. 1961. Effect of intermittant administration of a combination of amodiaquin and primaquine (Camoprim®) on the hematocrit of primaquine-sensitive and non-sensitive children. *American Journal of Tropical Medicine and Hygiene* 10(2): 128–134.

Holec, P., and R. J. Emry. 2003. Another molar of the Miocene hominid *Griphopithecus suessi* from the type locality at Sandberg, Slovakia. *Bulletin of the American Museum of Natural History* 279(1): 625–631.

Honigsbaum, M., and M. Willcox. 2004. Cinchona. In *Traditional medicinal plants and malaria*, M. Willcox, G. Bodeker, and Rasoanaivo, Eds., 21–41. Boca Raton, FL: CRC Press.

Horai, S., Y. Satta, K. Hayasaka, R. Kondo, T. Inoue, T. Ishida, S. Hayashi, and N. Takahata. 1992. Man's place in hominoidea revealed by mitochondrial DNA genealogy. *Journal of Molecular Evolution* 35(1): 32–43.

Hornberger, N. H. 1998. Language policy, language education, language rights: Indigenous immigrant, and international perspectives. *Language in Society* 27: 439–458.

Hornblum, A. M. 1997. They were cheap and available: Prisoners as research subjects in twentieth-century America. *British Medical Journal* 315(7120): 1437–1441.

Howard, L. M., and B. D. Cabrera. 1961. Simian malaria in the Philippines. *Science* 134(3478): 555–556.

Huang, W., R. Ciochon, G. Yumin, R. Larick, F. Qiren, H. Schwarcz, C. Yonge, J. de Vos, and W. Rink. 1995. Early *Homo* and associated artefacts from Asia. *Nature* 378(6554): 275–278.

Hughes, A. F., and F. Verra. 2010. Malaria parasite sequences from chimpanzee support the co-speciation hypothesis for the origin of virulent human malaria (*Plasmodium falciparum*). *Molecular Phylogenetics and Evolution* 57(1):135–143.

Huldén, L., L. Huldén, and K. Heliövaara. 2008. Natural relapses in *vivax* malaria induced by *Anopheles* mosquitoes. *Malaria Journal* 7: 64, *www.biomedcentral.com/content/pdf/1475-2875-7-64.pdf* (accessed August 4, 2010).

Humphreys, M. 2001. *Malaria: Poverty, race, and public health in the United States*. Baltimore: John Hopkins University Press.

Ijumba, J. N., and S. W. Lindsay. 2001. Impact of irrigation on malaria in Africa: Paddies paradox. *Medical and Veterinary Entomology* 15(1): 1–11.

International Union for Conservation of Nature and Natural Resources (IUCN) Red List of Threatened Species. 2009. *www.iucnredlist.org/* (accessed December 20, 2009).

Irie, A., S. Koyama, Y. Kozutsumi, T. Kawasaki, and A. Suzaki. 1998. The molecular basis for the absence of N-Glycolylneuraminic acid. *Journal of Biological Chemistry* 272(25): 15866–15871.

Jacob, T., E. Indriati, R. P. Soejono, K. Hsu, D. W. Frayer, R. B. Eckhardt, A. J. Kuperavage, A. Thorne, and M. Henneberg. 2006. Pygmoid Australomelanesian *Homo sapiens* skeletal remains from Liang Bua, Flores: Population affinities and pathological abnormalities. *Proceedings of the National Academy of Sciences, United States of America* 103(31): 13421–13426.

Jeffery, Geoffrey M. 1952. The infection of mosquitoes by *Plasmodium vivax* (Chesson strain) during the early parasitemias. *American Journal of Tropical Medicine and Hygiene* 1(4): 612–617.

———. 1956. Relapses with Chesson strain *Plasmodium vivax* following treatment with choloroquine. *American Journal of Tropical Medicine and Hygiene* 5(1): 1–13.

———. 1958. Infectivity to mosquitoes of *Plasmodium vivax* following treatment with cholorquine and other antimalarials. *American Journal of Tropical Medicine and Hygiene* 7(2): 207–211.

Jeffery, G. M., W. E. Collins, and J. C. Skinner. 1963. Antimalarial drug trials on a multiresistant strain of *Plasmodium falciparum*. *American Journal of Tropical Medicine and Hygiene* 12(6): 844–850.

Jeffery, G. M., and D. E. Eyles. 1954. The duration in the human host of infections with a Panama strain of *Plasmodium falciparum*. *American Journal of Tropical Medicine and Hygiene* 3(2): 219–224.

———. 1955. Infectivity to mosquitoes of *Plasmodium falciparum* as related to gametocyte density and duration of infection. *American Journal of Tropical Medicine and Hygiene* 4(5): 781–789.

Jeffery, G. M., G. B. Wolcott, M. D. Young, and D. Williams, Jr. 1952. Exo-erythrocytic stages of *Plasmodium falciparum*. *American Journal of Tropical Medicine and Hygiene* 1(6): 917–926.

Jeffery, G. M., M. D. Young, and A. Wilcox. 1954. The Donaldson strain of malaria. 1: History and characteristics of the infection in man. *American Journal of Tropical Medicine and Hygiene* 3(4): 628–637.

Jervis, H. R., H. Sprinz, A. J. Johnson, and B. T. Wellde. 1972. Experimental infection with *Plasmodium falciparum* in *Aotus* monkeys. II: Observations on host pathology. *American Journal of Tropical Medicine and Hygiene* 21(3): 272–281.

Jones, J. H. 1993. *Bad blood: The Tuskegee Syphilis Experiment*. New York: The Free Press.

Jongwutiwes, S., C. Putaporntip, T. Iwasaka, T. Sata, and H. Kanbara. 2004. Naturally acquired *Plasmodium knowlesi* malaria in humans, Thailand. *Emerging Infectious Diseases* 10(12): 2211–2213.

Joralemon, D. 1992. New World depopulation and the case of disease. *Journal of Anthropological Research* 38: 108–127.

Jukes, T. H. 1963. People and pesticides. *The American Scientist* 51(3): 355–361.

Kaifu, Y., H. Bab, F. Aziz, E. Indriati, F. Schrenk, and T. Jacob. 2005. Taxonomic affinities and evolutionary history of the early Pleistocene hominids of Java: Dentognathic evidence. *American Journal of Physical Anthropology* 128(4): 709–726.

Kantele, A., H. Marti, I. Felger, D. Müller, and T. S. Jokiranta. 2008. Monkey malaria in a European traveler returning from Malaysia. *Emerging Infectious Diseases* 14(9): 1434–1436.

Kantha, S. S. 1995. Is Karl Landsteiner the Einstein of the biomedical sciences? *Medical Hypotheses* 44(4): 254–256.

Kappelman, J., M. C. Alçiçek, N. Kazanc, M. Schultz, M. Özkul, and Ş. Şen. 2008. First *Homo erectus* from Turkey and implications for migrations into temperate Eurasia. *American Journal of Physical Anthropology* 135(1): 110–116.

Karafet, T. M., S. L. Zegura, and M. F. Hammer. 2006. Y chromosome variation. In *Environment, Origins, and Population*, D. H. Ubelaker, Ed., 817–830. *Handbook of North American Indians, Volume Three,* W. C. Sturtevant, Ed. Washington, D.C.: Smithsonian Institution Press.

Kaiser, M., A. Löwa, M. Ulrich, H. Ellerbrok, A. S. Goffe, A. Blasse, Z. Zommers, E. Couacy-Hymann, F. Babweteera, K. Zuberbühler, S. Metzger, S. Geidel, C. Boesch, T. R. Gillespie, and F. H. Leendertz. 2010. Wild chimpanzees infected with 5 *Plasmodium* species. *Emerging Infectious Diseases* 16(12): 1956–1959.

Kawai, S., M. Aikawa, S. Kano, and M. Suzuki. 1993. A primate model for severe human malaria: *Plasmodium coatneyi*-infected *Macaca fuscata. American Journal of Tropical Medicine and Hygiene* 48(5): 630–636.

Kawai, S., E. Ikeda, M. Sugiyama, J. Matsumoto, T. Higuchi, H. Zhang, N. Khan, K. Tomiyoshi, T. Inoue, H. Yamaguchi, K. Katakura, K. Endo, H. Matsuda, and M. Suzuki. 2006. Enhancement of splenic glucose metabolism during acute malaria infection: Correlation of finds of FDG-PET imaging with pathological changes in a primate model of severe human malaria. *American Journal of Tropical Medicine and Hygiene* 74(3): 353–360.

Kawai, S., S. Kano, and M. Suzuki. 1995. Rosette formation by *Plasmodium coatneyi*-infected erythrocytes of the Japanese macaque (*Macaca fuscata*). *American Journal of Tropical Medicine and Hygiene* 53: 295–299.

Kelly, D. W. 2001. Why are some people bitten more than others? *Trends in Parasitology* 17(12): 578–581.

Kelley, J. 2007. Identification of a single birth cohort in *Kenyapithecus kizili* and the nature of sympatry between *K. kizili* and *Griphopithcus alpani* at Palaşar. *Journal of Human Evolution* 54(4): 530–537.

Kelley, J., P. Andrews, and B. Alpagut. 2008. A new hominoid species from the middle Miocene site of Palaşar, Turkey. *Journal of Human Evolution* 54(4): 455–479.

Kelton, P. 2007. *Epidemics and enslavement: Biological catastrophe in the native Southeast, 1492–1715.* Lincoln: University of Nebraska Press.

King, N. W. 1994. The owl monkey in oncogenic virus research. In *Aotus: The owl monkey,* J. F. Baer, R. E. Weller, and I. Kakoma, Eds., 245–261. San Diego: Academic Press.

Kinzey, W. G., Ed. 1997. *New World primates: Ecology, evolution, and behavior.* New York: Aldine de Gruyter.

Kiple, K. F., and B. T. Higgins. 1992. Yellow fever and the africanization of the Caribbean. In *Disease and demography in the Americas*, J. W. Verano and D. H. Ubelaker, Eds., 237–248. Washington, D.C.: Smithsonian Institution Press.

Kirch, P. V. 2000. *On the road of the winds: An archaeological history of the Pacific Islands before European contact.* Berkeley and Los Angeles: University of California Press.

Klein, L. L., and D. J. Klein. 1976. Neotropical primates: Aspects of habitat usage, population density, and regional distribution in L Macerna, Colombia. In *Neotropical Primates: Field Studies and Conservation*, Eds., R. W. Thorington and P. B. Heltne, 70–79. Washington D.C.: National Academy of Sciences.

Klein, N., F. Fröhlich, and S. Krief. 2008. Geophagy: Soil consumption enhances the bioactivities of plants eaten by chimpanzees. *Naturwissenschaften* 95: 311–325.

Knobloch, J., J. D. Shaughnessy, Jr., and U. Rüther. 2007. Thalidomide induces limb deformities by perturbing the Bmp/Dkk1/Wnt signaling pathway. *FASEB Journal (Federation of American Societies for Experimental Biology)* 21: 1410–1421.

Knols, B. G. J., R. de Jong, and W. Takken. 1995. Differential attractiveness of isolated humans to mosquitoes in Tanzania. *Transactions of the Royal Society of Tropical Medicine and Hygiene* 89(6): 604–606.

Kohn, E. 2007. How dogs dream: Amazonian natures and the politics of transspecies engagement. *American Ethnologist* 34(1): 3–24.

Kordos, L. 2000. New results of hominoid research in the Carpathian Basin. *Acta Biologica Szegediensis* 44(1–4): 71–74.

Kordos, L., and D. R. Begun. 2001. Primates from Rudabánya: Allocation of specimens to individuals, sex, and age categories. *Journal of Human Evolution* 40(1): 17–35.

Krause. J., Q. Fu, J. M. Good, B. Viola, M. V. Shunkov, A. P. Derevianko, and S. Pääbo. 2010. The complete mitochondrial DNA genome of an unknown hominin from southern Siberia. *Nature* 464: 894–897.

Krief, S., M. A. Huffman, T. Sévent, C.-M. Hladik, P. Grellier, P. M. Loiseau, and R. W. Wrangham. 2006. Bioactive properties of plant species ingested by chimpanzees (*Pan troglyodytes schweinfurthii*) in the Kibale National Park, Uganda. *American Journal of Primatology* 68(1): 51–71.

Krief, S., J. M. Kasenene, M. Crandfield, O. E. Cornejo, J.-M. Chavatte, C. Lin, F. Letourneur, A. C. Grüner, T. F. McCutchan, L. Rénia, G. Snounou. 2010. On the diversity of malaria parasites in African apes and the origin of *Plasmodium falciparum* from Bonobos. *PLoSPathology* 6(2):e1000765, *www.plospathogens.org/article/info%3Adoi%2F10.1371%2Fjournal.ppat.1000765* (accessed January 24, 2011).

Kristensson, K. 1999. The discovery of the poliovirus. *Brain Research Bulletin* 50(5/6): 416.

Kroeber, K. 1997. Ecology and American literature: Thoreau and un-Thoreau. *American Literary History* 9(2): 309–328.

Kunimatsu, Y., B. Ratanasthien, H. Nakaya, H. Saegusa, and S. Nagaoka. 2003. Earliest Miocene hominoid from Southeast Asia. *American Journal of Physical Anthropology* 124(2): 99–108.

Kunimatsu, Y., B. Ratanasthien, H. Nakaya, H. Saegusa, S. Nagaoka, Y. Suganuma, A. Fukuchi, and B. Udomkan. 2005a. An additional specimen of a large-bodied Miocene hominoid from Chiang Muan, northern Thailand. *Primates* 46: 65–69.

———. 2005b. Hominoid fossils discovered from Chiang Muan, northern Thailand: The first step towards understanding hominoid evolution in Neogene Southeast Asia. *Anthropological Science* 113: 85–93.

Kusch, E., D. F. Milam, and W. K. Stratman-Thomas. 1936. General paresis treated by mosquito-inoculated *vivax* (tertian) malaria. *American Journal of Psychiatry* 93: 619–624.

Kuvin, S. F., H. K. Beye, F. Stohlman, Jr., P. G. Contacos, and G. R. Coatney. 1963. Malaria in man. Infection by *Plasmodium vivax* and the B strain of *Plasmodium cynomolgi*. *Journal of the American Medical Association* 184: 1018–1020.

Kuvin, S. F., J. E. Tobie, C. B. Evans, G. R. Coatney, and P. G. Contacos. 1962a. Flourescent antibody studies on the course of antibody production and serum gamma globulin levels in volunteers infected with human and simian malaria. *American Journal of Tropical Medicine and Hygiene* 11(4): 429–436.

———. 1962b. Antibody production in human malaria as determined by the flourse-cent antibody technique. *Science* 135(3509): 1130–1131.

Kwiatkowski, D. P. 2005. How malaria has affected the human genome and what human genetics can teach us about malaria. *American Journal of Human Genetics* 77(2): 171–192.

Lacroix, R., W. R. Mukabana, L. Clement Gouagna, and J. C. Koella. 2005. Malaria infection increases attractiveness of humans to mosquitoes. *PLoS Biology* 3(9): e298, *http://biology.plosjournals.org/archive/1545-7885/3/9/pdf/10.1371_journal.pbio.0030298-L.pdf* (accessed November 19, 2008).

Lal, A. A., V. F. de la Cruz, W. E. Collins, G. H. Campbell, P. M. Procell, and T. F. McCutchan. 1988. Circumsporozoite protein gene from *Plasmodium brasilianum*. Animal reservoirs for human malaria parasites? *The Journal of Biological Chemistry* 263: 5495–5498.

Lamin, B. S., U. Fillinger, C. Green, V. Louca, S. Majambere, and S. W. Lindsay. 2009. Agriculture and the promotion of insect pests: Rice cultivation in river floodplains and malaria vectors in The Gambia. *Malaria Journal* 8: 170, *www.malariajournal.com/content/8/1/170* (accessed August 8, 2010).

Lardeux, F., P. Loayza, B. Bouchité, and T. Chavez. 2007. Host choice and human blood index of *Anopheles pseudopunctipennis* in a village in the Andean valleys of Bolivia. *Malaria Journal* 6: 8, *www.lhl.uab.edu:15128/picrender.fcgi?tool=pmcentrez&artid=1783659&blobtype=pdf* (accessed May 31, 2008).

Larick, R., R. L. Ciochon, Y. Zaim, S. Suminto, Y. Rizal, F. Aziz, M. Reagan, and M. Heizler. 2001. Early Pleistocene $^{40}Ar/^{39}Ar$ ages for Bapang Formation hominins, Central Jawa, Indonesia. *Proceedings of the National Academy of Sciences, United States of America* 98(9): 4822–4871.

Latour, B. 2004. Politics of nature: How to bring the sciences into democracy. Cambridge, MA: Harvard University Press.

LeClerc, M. C., J. P. Hugot, P. Durand, and F. Renaud. 2004. Evolutionary relationships between 15 *Plasmodium* species from New and Old World primates (including humans): An 18s rDNA cladistic analysis. *Parasitology* 129: 677–684.

Lee, K.-S., J. Cox-Singh, G. Brooke, A. Matusop, and B. Singh. 2009. *Plasmodium knowlesi* from archival blood films: Further evidence that human infections are widely distributed and not newly emergent in Malaysian Borneo. *International Journal for Parasitology* 39(10): 1125–1128.

Lell, B., J. May, R. J. Schmidt Ott, L. G. Lehman, D. Luckner, B. Greve, P. Matousek, D. Schmid, K. Herbich, F. P. Mockenhaupt, C. G. Meyer, U. Bienzle, and P. G. Kremsner. 1999. The role of red blood cell polymorphisms in resistance and susceptibility to malaria. *Clinical Infectious Diseases* 28(4): 794–799.

Lesser, J. 1999. *Negotiating national identity: Immigrants, minorities, and the struggle for ethnicity in Brazil*. Durham, NC: Duke University Press.

Li, J., W. E. Collins, R. A. Wirtz, D. Rathore, A. Lal, and T. F. McCutchan. 2001. Geographic subdivision of the range of the malarial parasite *Plasmodium vivax*. *Emerging Infectious Diseases* 7: 35–42, *www.cdc.gov/ncidod/eid/vol7no1/li.htm* (accessed December 21, 2009).

Lim, C. S., L. Tazi, and F. J. Ayala. 2005. *Plasmodium vivax*: Recent world expansion and genetic identity to *Plasmodium simium. Proceedings of the National Academy of Science, United States of America* 102(43): 15523–15528.

Lindsay, S., J. Ansell, C. Selman, V. Cox, K. Hamilton, and G. Walraven. 2000. Effect of pregnancy on exposure to malaria mosquitoes. *Lancet* 355(9219): 1972.

Litsios, S. 2005. The health, poverty, and development merry-go-round: The tribulations of the WHO. In *Understanding the global dimensions of health*, Eds., S. W. A. Gunn, P. B. Mansourian, A. M. Davies, A. Piel, and B. McA. Sayers, 15–33. New York: Springer.

Liu, W., Y. Li, G. H. Learn, R. S. Rudicell, J. D. Robertson, B. F. Keele, J.-B. N. Ndjango, C. M. Sanz, D. B. Morgan, S. Locatelli, M. K. Gonder, P. J. Kranzusch, P. D. Walsh, E. Delaporte, E. Mpoudi-Ngole, A. V. Georgiev, M. N. Muller, G. M. Shaw, M. Peeters, P. M. Sharp, J. C. Rayner, and B. H. Hahn. 2010. Origin of the human malaria parasite *Plasmodium falciparum* in gorillas. *Nature* 467: 420–425.

Livingstone, F. B. 1958. Anthropological implications of sickle cell gene distribution in West Africa. *American Anthropologist* 60: 533–562.

———. 1971. Malaria and human polymorphisms. *Annual Review of Genetics* 5: 33–64.

———. 1984. The Duffy blood groups, *vivax* malaria, and malaria selection in human populations: A review. *Human Biology* 56(3): 413–425.

Loftin, H. 1955. Monkey business. *Science News-Letter* 68(15): 234–235.

Long, H. Y., B. Lell, K. Dietz, and P. G. Kremsner. 2001. *Plasmodium falciparum*: In vitro growth inhibition by febrile temperatures. *Parasitology Research* 87(7): 553–555.

Lordkipanidze, D., A. Vekua, R. Ferring, G. P. Rightmire, C. P. E. Zollikofer, M. S. Ponce de León, J. Agusti, G. Kiladze, A. Mouskhelishvili, M. Nioradze, and M. Tappen. 2006. A fourth hominin skull from Dmanisi, Georgia. *The Anatomical Record Part A: Discoveries in Molecular, Cellular, and Evolutionary Biology* 288A(11): 1146–1157.

Loscertales, M.-P., and B. J. Brabin. 2006. ABO phenotypes and malaria related outcomes in mothers and babies in the Gambia: A role for histo-blood groups in placental malaria? *Malaria Journal* 5:72, *www.malariajournal.com/content/pdf/1475-2875-5-72.pdf* (accessed May 31, 2008).

Lourenço-de-Oliveira, R., and L. M. Deane. 1995. Simian malaria at two sites in the Brazilian Amazon: I—The infection rates of *Plasmodium brasilianum* in non-human primates. *Memórias do Instituto Oswaldo Cruz* 90(3) (May/June), *www.bioline.org.br/request?oc95066* (accessed September 3, 2009).

Luchavez, J., F. Espino, and P. Curameng. 2008. Human infections with *Plasmodium knowlesi*, the Philippines. *Emerging Infectious Diseases* 14(5): 811–813.

Lunn, J. F., W. Chin, P. G. Contacos, and G. R. Coatney. 1964. Cycloguanil pamoate (CI-501) as a causal prophylactic against a southern Rhodesian strain of *falciparum* malaria. *American Journal of Tropical Medicine and Hygiene* 13(6): 783–785.

———. 1966. Changes in antibody titers and serum protein fractions during the course of prolonged infections with *vivax* or with *falciparum* malaria. *American Journal of Tropical Medicine and Hygiene* 15(1): 3–10.

Lunn, J. F., R. L. Jacobs, P. G. Contacos, and G. R. Coatney. 1965. Antibody production in *Plasmodium vivax* infections suppressed by weekly doses of chloroquine. *American Journal of Tropical Medicine and Hygiene* 14(5): 697–699.

Luzzatto, L. 2006. Glucose 6-phosphate dehydrogenase deficiency: From genotype to phenotype. *Haematologica* 91(10): 1303–1306.

Luzzatto, L., E. S. Nwachiku-Jarrett, and S. Reddy. 1970. Increased sickling of parasitised erythrocytes as a mechanism of resistance against malaria in sickle cell trait. *Lancet* 1(7642): 319–321.

MacArthur, R. H., and E. R. Pianka. 1966. On the optimal use of a patchy environment. *American Naturalist* 100(916): 603–609.

MacFie, T. S., E. Nerrienet, R. E. Bontrop, and N. I. Mundy. 2009. The action of *falciparum* malaria on the human and chimpanzee genomes compared: Absence of evidence for a genomic signature of malaria at HBB and G6PD in three subspecies of chimpanzee. *Infection, Genetics, and Evolution* 9(6): 1248–1252.

Madar, S. I., M. D. Rose, J. Kelley, L. MacLatchy, and D. Pilbeam. 2002. New *Sivapithecus* postcranial specimens from the Siwaliks of Pakistan. *Journal of Human Evolution* 42(6): 705–752.

Maeno, Y., A. E. Brown, C. D. Smith, T. Tegoshi, T. Toyoshima, C. F. Ockenhouse, K. D. Corcoran, M. Ngampochjana, D. E. Kyle, H. K. Webster, and M. Aikawa. 1993. A nonhuman primate model for human cerebral malaria: Effects of artesunate (quinghaosu deriviative) on Rhesus monkeys experimentally infected with *Plasmodium coatneyi. American Journal of Tropical Medicine and Hygiene* 49(6): 726–734.

Maier, A. G., M. T. Durasingh, J. C. Reeder, S. S. Patel, J. W. Kazura, P. A. Zimmerman, and A. F. Cowman. 2002. *Plasmodium falciparum* erthrocyte invasion through glycophorin C and selection for Gerbich negativity in human populations. *Nature* 9: 87–92.

Marchant, A. 1961. Dom Joao's botanical garden. *The Hispanic Historical Review* 41: 259–274.

Marley, F. 1963. Monkeys stand in for humans. *Science News-Letter* 84(14): 218–219.

Marrama, L., R. Jambou, I. Rakotoarivony, J. M. Leong Pock Tsi, J. B. Duchemin, S. Laventure, J. Mouchet, and J. Roux. 2004. Malaria transmission in Southern Madagascar: Influence of the environment and hydro-agricultural works in subarid and humid regions Part 1. Entomological investigations. *Acta Tropica* 89: 193–203.

Marques, A. C. 1987. Human migration and the spread of malaria in Brazil. *Parasitology Today* 3(6): 166–170.

Marsh, K. 2002. Immunology of malaria. In *Essential malariology*, fourth edition, D. A. Warell and H. M. Gilles, Eds., 252–267. New York: Oxford University Press.

Martin, M. J., J. Rayner, P. Gagneux, J. W. Barnwell, and A. Varki. 2005. Evolution of human-chimpanzee differences in malaria susceptibility: Relationship to human genetic loss of *N*-glycolylneuraminic acid. *Proceedings of the National Academy of Science, United States of America* 102(36): 12819–12824.

Martin, R. D., A. M. Maclarnon, J. L. Phillips, L. Dussubieux, P. R. Williams, and W. B. Dobyns. 2006a. Comment on "The brain of LB1, *Homo floresiensis.*" *Science* 312(5776): 999b.

———. 2006b. Flores hominid: New species or microcephalic dwarf? *The Anatomical Record Part A* 288A(11): 1123–1145.

Martinsen, E. S., S. L. Perkins, and J. J. Schall. 2008. A three-genome phylogeny of malaria parasites (*Plasmodium* and closely related genera): Evolution of life-history traits and host-switches. *Molecular Phylogenetics and Evolution* 47(1): 261–273.

Matsumoto, J., S. Kawai, K. Terao, M. Kirinoki, Y. Yasutomi, M. Aikawa, and H. Matsuda. 2000. Malaria infection induces rapid elevation of soluble fas ligand level in serum and subsequent T lymphocytopenia: Possible factors responsible for the differences in susceptibility of two species of *Macaca* monkeys to *Plasmodium coatneyi* infection. *Infection and Immunity* 68(3): 1183–1188.

Matteson, P. 1999. *Resolving the DDT dilemma: Protecting biodiversity and human health*. Washington, D.C.: World Wildlife Fund.

Matthys, B., T. Sherkanov, S. S. Karimov, Z. Khabirov, T. Mostowlansky, J. Utzinger, and K. Wyss. 2008. History of malaria control in Tajikistan and rapid malaria appraisal in an agro-ecological setting. *Malaria Journal 7:217, www.malariajournal.com/content/7/1/217* (accessed August 8, 2010).

May, J., J. A. Evans, C. Timmann, C. Ehmen, W. Busch, T. Thye, T. Agbenyega, and R. D. Horstmann. 2007. Hemoglobin variants and disease manifestation in severe *falciparum* malaria. *Journal of the American Medical Association* 297(20): 2220–2226.

Mayr, E., and R. J. O'Hara. 1986. The biogeographic evidence supporting the Pleistocene forest refuge hypothesis. *Evolution* 40(1): 55–67.

McCauley, C. 2005. Aggressive active case detection: A malaria control strategy based on the Brazilian model. *Social Science & Medicine* 60: 563–573.

McCutchen, T. F. 2008. Is monkey malaria from Borneo an emerging human disease? *Future Microbiology* 3(2): 115–118.

McCutchan, T. F., J. B. Dame, L. H. Miller, and J. Barnwell. 1984. Evolutionary relatedness of *Plasmodium* species as determined by the structure of DNA. *Science* 225(4664): 808–811.

McCutchan, T. F., J. C. Kissenger, M. G. Touray, M. J. Rogers, J. Li, M. Sullivan, E. M. Braga, A. U. Krettli, and L. H. Miller. 1996. Comparison of circumsporozoite proteins from avian and mammalian malarias: Biological and phylogenetic implications. *Proceedings from the National Academy of Sciences, United States of America* 93(21): 11889–11894.

McGovern, T. H. 1990. The archaeology of the Norse North Atlantic. *Annual Review of Anthropology* 19: 331–351.

McGraw, W. Scott, and John G. Fleagle. 2006. Biogeography and evolution of the *Cercocebus-Mandrillus* Clade: Evidence from the face. In *Primate biogeography: Progress and prospects*, S. M. Lehman and J. G. Fleagle, Eds., 201–224. New York: Springer.

McNeill, W. H. 1976. *Plagues and peoples*. Garden City, NY: Anchor.

Meagher, E. T. 1929. *General paralysis and its treatment by induced malaria*. London: H. M. Stationary Office.

Meggers, B. J. 1957. Environment, history, and culture in the Amazon basin: An appraisal of the theory of environmental determinism. *Studies in Human Ecology, Social Science Monographs* No. 111, 71–89. Washington, D.C.: Pan American Union.

———. 1973. Some problems of cultural adaptation in Amazonia, with emphasis on the pre-European period. In *Tropical forest ecosystems in Africa and South America: A comparative review*, B. J. Meggers, E. S. Ayensu, and W. D. Duckworth, Eds., 311–320. Washington, D.C.: Smithsonian Institution Press.

Mendis, K., B. J. Sina, P. Marchesini, and R. Carter. 2001. The neglected burden of *Plasmodium vivax* malaria. *American Journal of Tropical Medicine and Hygiene* 64: 97–106.

Merriwether, D. A. 2006. Mitochondrial DNA. In *Environment, Origins, and Population*, D. H. Ubelaker, Ed., 817–830. *Handbook of North American*

Indiians, Volume Three, W. C. Sturtevant, Ed. Washington, D.C.: Smithsonian Institution Press.

Metcalf, R. L. 1973. A century of DDT. *Journal of Agriculture and Food Chemistry* 21(4): 511–519.

Meuwissen, J. H. E. T. 1961. Resistance of *Plasmodium falciparum* to pyrimethamine and proguanil in Netherlands, New Guinea. *American Journal of Tropical Medicine and Hygiene* 10(2): 135–139.

Migot-Nabias, F., B. Ollomo, G. Dubreuil, A. Morelli, O. Domarle, R. Nabias, and A. J. Georges. 1999. *Plasmodium coatneyi:* Differential clinical and immune responses of two populations of *Macaca fasicularis* from different origins. *Experimental Parasitology* 91: 30–39.

Miller, L. H. 1969. Distribution of mature trophozoites and schizonts of *Plasmodium falciparum* in the organs of *Aotus trivirgatus*, the night monkey. *American Journal of Tropical Medicine and Hygiene* 18(6): 860–865.

Miller, L. H., S. Chien, and S. Usami. 1972. Decreased deformability of *Plasmodium coatneyi*-infected red cells and its possible relation to cerebral malaria. *American Journal of Tropical Medicine and Hygiene* 21(2): 133–137.

Miller, L. H., and H. N. Fremount. 1970. Sequestration of schizonts of *Plasmodium coatney* in adipose tissue. *Journal of Parasitology* 56(5): 1028–1029.

Miller, L. H., H. N. Fremount, and S. A. Luse. 1971. Deep vascular schizogony of *Plasmodium knowlesi* in *Macaca mulatta. American Journal of Tropical Medicine and Hygiene* 20(6): 816–824.

Miller, L. H., M. F. Good, and G. Milon. 1994. Malaria pathogenesis. *Science* 264(5167): 1878–1883.

Miller, L. H., D. J. Wyler, R. H. Glew, W. E. Collins, and P. G. Contacos. 1974. Sensitivity of four Central American strains of *Plasmodium vivax* to primaquine. *American Journal of Tropical Medicine and Hygiene* 23(2): 309–310.

Miller, L. H., S. J. Mason, D. F. Clyde, and M. H. McGinniss. 1976. The resistance factor to *Plasmodium vivax* in blacks. The Duffy-blood-group genyotype, FyFy. *New England Journal of Medicine*. 295(6): 302–304.

Millet, P., T. L. Fisk, W. E. Collins, J. R. Broderson, and P. Nguyen-Dinh. 1988. Cultivation of exoerythrocytic stages of *Plasmodium cynomolgi, P. knowlesi, P. coatneyi*, and *P. inui* in *Macaca mulatta* hepatocytes. *Journal of Tropical Medicine and Hygiene* 39(6): 529–534.

Ming-Chu Kuo, M.-C., T.-Y. Chiang, C.-W. Chan, W.-S. Tsai, and D.-D. Ji. 2009. A case report of simian malaria, *Plasmodium knowlesi*, in a Taiwanese traveler from Palawan Island, the Philippines. *Epidemiology Bulletin* 25(3): 178–191.

Modiano, D., G. Luoni, B. S. Sirima, J. Simporé, F. Verra, A. Konaté, E. Rastrelli, A. Oliveira, C. Calissano, G. M. Paganotti, L. D'Urbano, I. Sanou, A. Sawadogo, G. Modiano, and M. Couzzli. 2001. Haemoglobin C protects against clinical *Plasmodium falciparum* malaria. *Nature* 414: 305–308.

Montenegro, A., C. Avis, and A. Weaver. 2008. Modeling the prehistoric arrival of the sweet potato in Polynesia. *Journal of Archaeological Science* 35(2): 355–367.

Moreno, A., A. García, M. Cabrera-Mora, E. Strobert, and M. R. Galinski. 2007. Case report: Disseminated intravascular coagulation complicated by peripheral gangrene in a Rhesus macaque (*Macaca mulatta*) experimentally infected with *Plasmodium coatneyi. American Journal of Tropical Medicine and Hygiene* 76(4): 648–654.

Morris, C. L., J. S. Sullivan, H. M. McClure, E. Strobert, B. B. Richardson, G. G. Galland, I. F. Goldman, and W. E. Collins. 1996. The Nigerian I/CDC strain of *Plasmodium ovale* in chimpanzees. *Journal of Parasitology* 82(3): 444–448.

Morwood, M. J., P. Brown, Jatmiko, T. Sutikna, E. W. Saptomo, K. E. Westaway, R. A. Due, R. G. Roberts, T. Maeda, S. Wasisto, and T. Dubiantono. 2005. Further evidence for small-bodied hominins from the Late Pleistocene of Flores, Indonesia. *Nature* 437(7061): 957–958.

Moyà-Solà, S., D. M. Alba, S. Almécija, I. Casanovas-Vilar, M. Köhler, S. de Esteban-Trivigno, J. M. Robles, J. Galindo, and J. Fortuny. 2009. A unique middle Miocene European hominoid and the origins of the great ape and human clade. *Proceedings of the National Academy of Science, United States of America* 106(24): 9601–9606.

Moyà-Solà, S., and M. Köhler. 1993. Recent discoveries of *Dryopithecus* shed light on the evolution of great apes. *Nature* 365(6446): 543–545.

Moyà-Solà, S., M. Köhler, and D. M. Alba. 2001. *Egarapithecus narcisoi*, a new genus of Pliopithecidae (Primates, Catarrhini) from the late Miocene of Spain. *American Journal of Primatology* 114(4): 17–35.

Moynihan, M. 1976. *The New World primates: Adaptive radiation and the evolution of social behavior, languages, and intelligence*. Princeton, NJ: Princeton University Press.

Mu, J., J. Duan, K. D. Makova, D. A. Joy, C. Q. Huynh, O. H. Branch, W.-H. Li, and X.-Z. Su. 2002. Chromosome-wide SNPs reveal an ancient origin for *Plasmodium falciparum*. *Nature* 418: 323–326.

Mu, J., D. A. Joy, J. Duan, Y. Huang, J. Carlton, J. Walker, J. Barnwell, P. Beerli, M. A. Charleston, O. G. Pybus, and X.-Z. Su. 2005. Host switch leads to emergence of *Plasmodium vivax* malaria in humans. *Molecular and Biological Evolution* 22(8): 1686–1693.

Muchmore, E. A., S. Diaz, and A. Varki. 1998. A structural difference between the cell surfaces of humans and the great apes. *American Journal of Physical Anthropology* 107(2): 187–198.

Mueller, I., P. A. Zimmerman, and J. C. Reeder. 2007. *Plasmodium malariae* and *Plasmodium ovale* — the 'bashful' malaria parasites. *Trends in Parasitology* 23: 278–283.

Muirhead-Thomson, R. C. 1951. The distribution of anopheline mosquito bites among different age groups: A new factor in malaria epidemiology. *British Medical Journal* 1(4715): 1114–1117.

Mukabana, W. R., W. Takken, R. Coe, and B. G. J. Knols. 2002. Host-specific cues cause differential attractiveness of Kenyan man to African malaria vector *Anopheles gambiae*. *Malaria Journal* 1:1, *www.lhl.uab.edu:15128/picrender.fcgi?tool=pmcentrez&artid=149388&blobtype=pdf* (accessed May 30, 2008).

Mukabana, W. R., W. Takken, G. F. Killen, and B. G. J. Knols. 2004. Allomonal effect of breath contributes to differential attractiveness of humans to the African malarial vector *Anophels gambiae*. *Malaria Journal* 3:1, *www.lhl.uab.edu:15128/picrender.fcgi?tool=pmcentrez&artid=343289&blobtype=pdf* (accessed May 31, 2008).

Munyao, J. M., M. M. Gicheru, Z. W. Ng'ang'a, E. Kagasi, and H. S. Ozwara. 2008. Transinfection of *Plasmodium knowlesi* in baboon. *International Journal of Integrative Biology* 2(2): 100–108.

Muriu, S. M., E. J. Muturi, J. I. Shililu, C. M. Mbogo, J. M. Mwangangi, B. G. Jacob, L. W. Irungu, R. W. Mukabana, J. I. Githure, and R. J. Novak. 2008. Host choice and multiple blood feeding behaviour of malaria vectors and other anophelines in Mwea Rice Scheme, Kenya. *Malaria Journal* 7: 43, *www.malariajournal.com/content/pdf/1475-2875-7-43.pdf* (accessed May 31, 2008).

Myatt, A. V., G. R. Coatney, T. Hernandez, and H. W. Burton. 1953. A further study on the toxicity of pyrimethamine (daraprim) in man. *American Journal of Tropical Medicine and Hygiene* 2(6): 1000–1001.

Myatt, A. V., T. Hernandez, and G. R. Coatney. 1953. Studies in human malaria. XXXIII: The toxcitiy of pyrimethamine (daraprim) in man. *American Journal of Tropical Medicine and Hygiene* 2(5): 788.

Nacher, M. 2005. Charming the mosquito: Do malaria symptoms increase the attractiveness of the host for the vector? *Medical Hypotheses* 64(4): 788–791.

Nakano, Y., H. Fujioka, K. D. Luc, J. R. Rabbege, G. D. Todd, W. E. Collins, and M. Aikawa. 1996. A correlation of the sequestration rate of *Plasmodium coatneyi*-infected erythrocytes in cerebral and subcutaneous tissues of a Rhesus monkey. *American Journal of Tropical Medicine and Hygiene* 55(3): 311–314.

Nargolwalla, M. C., D. R. Begun, M. C. Dean, D. J. Reid, and L. Kordos. 2005. Dental development and life history in *Anapithecus hernyaki. Journal of Human Evolution* 49(1): 99–121.

National Institute of Neurological Disorders and Stroke. 2009. NINDS Neurosyphilis Information Page, *www.ninds.nih.gov/disorders/neurosyphilis/neurosyphilis.htm* (accessed August 4, 2010).

Nkhoma, E. T., C. Poole, V. Vannappagari, S. A. Hall, and E. Beutler. 2009. The global prevalence of glucose-6-phosphate dehydrogenase deficiency: A systematic review and meta-analysis. *Blood Cells, Molecules, and Diseases* 42(3): 267–278.

O'Brien, P. 1972. The sweet potato: Its origin and dispersal. *American Anthropologist* 74(3): 342–365.

Ohashi, J., I. Naka, J. Patarapotikul, H. Hananantachai, G. Brittenham, S. Looareesuwan, A. G. Clark, and K. Tokunaga. 2004. Extended linkage disequilibrium surrounding the hemoglobin E variant due to malarial selection. *American Journal of Human Genetics* 74(6): 1198–1208.

Oladepo, O., G. O. Tona, F. O. Oshiname, and M. A. Titiloye. 2010. Malaria knowledge and agricultural practices that promote mosquito breeding in two rural farming communities in Oyo State, Nigeria. *Malaria Journal 9:9, www.malariajournal.com/content/9/1/91* (accessed August 5, 2010).

O'Leary, D. S., C. F. Barr, B. T. Wellde, and M. E. Conrad. 1972. Experimental infection with *Plasmodium falciparum* in *Aotus* monkeys. III: The development of disseminated intravasular coagulation. *American Journal of Tropical Medicine and Hygiene* 21(3): 282–287.

Ollomo, B., P. Durand, F. Prugnolle, E. Douzery, C. Arnathau, D. Nkoghe, E. Leroy, and F. Renaud. 2009. A new malaria agent in African hominids. *PloS, Pathology* May 5(5), *www.plospathogens.org/article/info%3Adoi%2F10.1371%2Fjournal.ppat.1000446* (accessed September 9, 2009).

Ollomo, B., S. Karch, P. Bureau, N. Elissa, A. J. Georges, and P. Millet. 1997. Lack of malarial parasite transmission between apes and humans in Gabon. *American Journal of Tropical Medicine and Hygiene* 56(4): 440–445.

Oppenheimer, J. R. 1977. *Presbytis entellus*, the Hanuman languar. In *Primate conservation*, His Serene Highness Prince Ranier III of Monaco and G. H. Bourne, Eds., 469–512. New York: Academic Press.

Otsuru, M., and H. Sekikawa. 1968. A survey of simian malaria in Japan. *Transactions of the Royal Society of Tropical Medicine and Hygiene* 62(4): 558–561.

Ozwara, H., J. A. M. Langermans, J. Maamun, I. O. Farah, D. S. Yole, J. M. Mwenda, H. Weiler, and A. W. Thomas. 2003. Experimental infection of the olive baboon (*Papio anubis*) with *Plasmodium knowlesi*: Severe disease accompanied by cerebral involvement. *American Journal of Tropical Medicine and Hygiene* 69(2): 188–194.

Page, R. D. M., and M. A. Charleston. 1998. Trees within trees: Phylogeny and historical associations. *Trends in Ecology and Evolution* 13(9): 356–359.

Palatnik, M., and A. W. Rowe. 1984. Duffy and duffy-related human antigens in primates. *Journal of Human Evolution* 13: 173–179.

Palmieri, J., R. Purnomo, V. H. Lee, D. T. Dennis, and H. A. Marwoto. 1980. Parasites of the Silvered Leaf Monkey, *Presbytis cristatus* Eschscholtz 1921, with a note on wuchereria-like nematode. *Journal of Parasitology* 66(1): 170–171.

Pan American Health Organization. 2006. Health of indigenous populations in the Americas: Report of the ad-hoc consultative group on the health agenda for the Americas. *138th Session of the Executive Committee*. Washington, D.C., *www.paho.org/English/GOV/CE/ce138-inf5-e.pdf* (accessed February 15, 2008).

Pan, R., C. Groves, and C. Oxnard. 2004. Relationships between the fossil Colobine *Mesopithecus pentelicus* and extant Cercopithecoids, based on dental metrics. *Journal of Human Evolution* 62(4): 287–299.

Paris, Q. 1992. The von Liebig hypothesis. *American Journal of Agricultural Economics* 74(4): 1019–1028.

Patterson, N., D. J. Richter, S. Gnerre, E. S. Lander, and D. Reich. 2006. Genetic evidence for complex speciation of humans and chimpanzees. *Nature* 441: 1103–1108.

Perkins, S. L., and J. J. Schall. 2002. A molecular phylogeny of malarial parasites recovered from cytochrome *b* gene sequences. *Journal of Parasitology* 88(5): 972–978.

Peters, C. J., E. D. Johnson, P. B. Jahrling, T. G. Ksiazek, P. E. Rollin, J. White, W. Hall, R. Trotter, and N. Jaax. 1993. Filoviruses. In *Emerging Viruses*, Stephen S. Morse, Ed., 159–175. New York: Oxford University Press.

Peters, W., P. C. C. Garnham, R. Killick-Kendrick, N. Rajapaksa, W. H. Cheong, and F. C. Cadigan. 1976. Malaria of the Orangutan. *Philosophical Transactions of the Royal Society of London. Series B. Biological Sciences* 275(941): 439–482.

Pilbeam, D., M. D. Rose, J. C. Barry, and S. M. Ibrahim Shah. 1990. New *Sivapithecus humeri* from Pakistan and the relationship of *Sivapithecus* and *Pongo*. *Nature* 348(6298): 237–239.

Poirriez, J., E. DeiCas, L. Dujardin, and I. Landau. 1995. The blood-stages of *Plasmodium georgesi*, *P. gonderi* and *P. petersi*: Course of untreated infection in their natural hosts and additional morphological distinctive features. *Parasitology* 111: 547–554.

Poirriez, J., E. Dei-Cas, and I. Landau. 1994. Further description of the blood stages of *Plasmodium petersi* from *Cercocebus albigena* monkey. *Folia Parasitologica* 41: 148–182.

Pollet, H., and C. F. Barr. 1968. DNA, RNA, and protein synthesis in erythrocytic forms of *Plasmodium knowlesi*. *American Journal of Tropical Medicine and Hygiene* 17(5): 672–679.

Porter, J. A., Jr., 1971. *Plasmodium vivax* infections in *Aotus trivirgatus*. *American Journal of Tropical Medicine and Hygiene* 20(4): 535–538.

Porter, J. A., Jr., C. M. Johnson, and L. De Sousa. 1966. Prevalence of malaria in Panamanian primates. *Journal of Parasitology* 52(4): 669–670.

Porter, J. A., Jr., and M. D. Young. 1967. The transfer of *Plasmodium falciparum* from man to the marmoset, *Saguinus geoffroyi*. *Journal of Parasitology* 53(4): 845–846.

———. 1970. *Plasmodium vivax* infections in the spider monkeys, *Ateles fusciceps* and *Ateles geoffroyi*. *Journal of Parasitology* 56(3): 426–430.

Powell, R. D., and G. J. Brewer. 1965. Glucose-6-phosphate dehydrogenase deficiency and *falciparum* malaria. *American Journal of Tropical Medicine and Hygiene* 14(3): 358–362.

———. 1967. Effects of pyrimethamine, chlorguanide, and primatquine against exoerythrocytic forms of a strain of cholorquine-resistant *Plasmodium falicparum* from Thailand. *American Journal of Tropical Medicine and Hygiene* 16(6): 693–698.

Powell, R. D., G. J. Brewer, and A. S. Alving. 1963. Studies of a strain of choloquine-resistant *Plasmodium falciparum* from Colombia, South America. *American Journal of Tropical Medicine and Hygiene* 12(4): 509–512.

Powell, R. D., R. L. DeGowin, and R. B. Eppes. 1965. Studies on the antimalarial effects of cycloguanil pamoate (CI-501) in man. *American Journal of Tropical Medicine and Hygiene* 14(6): 913–921.

Price, R. N., E. Tjitra, C. A. Guerra, S. Yeung, N. J. White, and N. M. Anstey. 2007. *Vivax* malaria: Neglected and not benign. *American Journal of Tropical Medicine and Hygiene* 77(6): 79–87.

Prugnolle, P. D., C. Neel, B. Ollomo, F. J. Ayala, C. Arnathau, L. Etienne, E. Mpoudi-Ngole, D. Nkoghe, E. Leroy, E. Delaporte, M. Peeters, and F. Renaud. 2010. African great apes are natural hosts of multiple related malaria species, including *Plasmodium falciparum*. *Proceedings of the National Academy of Science,United States of America* 107(4): 1458–1463.

Prugnolle, P. D., C. Neel, B. Ollomo, F. J. Ayala, C. Arnathau, L. Etienne, E. Mpoudi-Ngole, D. Nkoghe, E. Leroy, E. Delaporte, M. Peeters, and F. Renaud. 2010. African great apes are natural hosts of multiple related malaria species, including *Plasmodium falciparum*. *Proceedings of the National Academy of Sciences, United States of America* 107(4): 1458–1463.

Puri, S. K., and G. P. Dutta. 1994. *Presbytis entellus*, a new experimental host for *Plasmodium fragile*. *Journal of Parasitology* 80(1): 156–158.

Putaporntip, C., T. Hongsrimuang, S. Seethamchai, T. Kobasa, K. Limkittikul, L. Cui, and S. Jongwutiwes. 2009. Differential prevalence of *Plasmodium* infections and cryptic *Plasmodium knowlesi* malaria in humans in Thailand. *Journal of Infectious Diseases* 199(8): 1143–1150.

Putaporntip, C., S. Jongwuitwes, S. Thongaree, S. Seethamchai, P. Gyrnberg, and A. L. Hughes. 2010. Ecology of malaria parasites infecting Southeast Asian macaques: Evidence from cytochrome *b* sequences. *Molecular Ecology* 19: 3466–3476.

Qari, S. H., Y. P. Shi, N. J. Pieniazek, W. E. Collins, and A. A. Lal. 1996. Phylogenetic relationship among the malaria parasites based on small subunit rRNA gene sequences: Monophyletic nature of the human malaria parasite *Plasmodium falciparum*. *Molecular Phylogenetics and Evolution* 6(1): 157–165.

Raaum, R. L., K. N. Sterner, C. M. Noviello, C.-B. Stewart, and T. R. Disotell. 2005. Catarrhine primate divergence dates estimated from complete mitochondrial genomes: Concordance with fossil and nuclear DNA evidence. *Journal of Human Evolution* 48(3): 237–257.

Rabinow, P. 1996. Artificiality and enlightenment: From sociobiology to biosociality. In *Essays on the anthropology of reason*, 91–111. Princeton, NJ: Princeton University Press.

Rajkumar, S. V. 2004. Thalidomide: Tragic past and promising future. *Mayo Clinic Proceedings* 79(7): 899–903.

Ramenofsky, A. 1993. Diseases of the Americas, 1492–1700. In *The Cambridge World History of Human Disease*, K. F. Kiple, Ed., 317–327. Cambridge: Cambridge University Press.

Ranson, H., B. Jensen, J. M. Vulule, X. Wang, J. Hemingway, and F. H. Collins. 2000. Identification of a point mutation in the voltage-gated sodium channel gene of Kenyan *Anopheles gambiae* associated with resistance to DDT and pyrethroids. *Insect Molecular Biology* 9(5): 491–497.

Reid, M. J. C., R. Ursic, D. Cooper, H. Nazzari, M. Griffiths, B. M. Galdikas, R. M. Garriga, M. Skinner, and C. Lowenberger. 2006. Transmission of human and macaque *Plasmodium* spp. to ex-captive orangutans in Kalimantan, Indonesia. *Emerging Infectious Diseases* 12(12) (December), *www.cdc.gov/ncidod/eid/vol12no12/pdfs/06-0191.pdf* (accessed June 13, 2008).

Reiter, P. 2008. Global warming and malaria: Knowing the horse before hitching the cart. *Malaria Journal* 2008, 7(Suppl 1): S3, *www.malariajournal.com/content/7/S1/S3* (accessed August 5, 2010).

Renquist, D. M., and R. A. Whitney, Jr. 1987. Zoonoses acquired from pet primates. *Veterinary Clinics of North America: Small Animal Practice* 17: 219–240.

Rich, S. M. 2004. The unpredictable past of *Plasmodium vivax* revealed in its genome. *Proceedings of the National Academy of Science, United States of America* 101(44): 15547–15548.

Rich, S. M., and F. J. Ayala. 2000. Population structure and recent evolution of *Plasmodium falciparum. Proceedings of the National Academy of Sciences, United States of America* 97(13): 6994–7001.

Rich, S. M., F. H. Leendertz, G. Xu, M. Lebreton, C. F. Djoko, M. N. Aminake, E. E. Takang, J. L. Diffo, B. L. Pike, B. M. Rosenthal, P. Formenty, C. Boesch, F. J. Ayala, and N. D. Wolf. 2009. The origin of malignant malaria. *Proceedings of the National Academy of Science, United States of America* 106(35): 14902–14907.

Rich, S. M., M. C. Light, R. R. Hudson, and F. J. Ayala. 1998. Malaria's Eve: Evidence of a recent population bottleneck throughout the world populations of *Plasmodium falciparum. Proceedings of the National Academy of Sciences, United States of America* 95(8): 4425–4430.

Ricklefs, R., and D. C. Outlaw. 2010. A molecular clock for malaria parasites. *Science* 329 (5988): 226–229.

Rieckmann, K. H., R. D. Powell, J. V. McNamara, D. Willerson, Jr., L. Kass, H. Frischer, and P. E. Carson. 1971. Effects of tetracycline against chloroquine-resistant and cholorquine-sensitive *Plasmodium falciparum. American Journal of Tropical Medicine and Hygiene* 20(6): 811–815.

Rigdon, R. H., and W. K. Stratman Thomas. 1942. A study in the pathological lesions in *P. knowlesi* infection in M. Rhesus monkeys. *American Journal of Tropical Medicine and Hygiene* s1-22(4): 329–339.

Rodhain, J. 1948. Susceptibility of the chimpanzee to *P. malariae* of human origin. *American Journal of Tropical Medicine and Hygiene* s1-28(5): 629–631.

Rook, L. 1999. Late Turolian *Mesopithecus* (Mammalia, Primates, Colobinae) from Italy. *Journal of Human Evolution* 36(5): 535–594.

Rook, L., and P. O'Higgins. 2005. A comparative study of adult facial morphology and its ontogeny in the fossil macaque *Macaca majori* from the Capo Figari, Sardina, Italy. *Folia Primatologia* 76(3): 151–171.

Rook, L., A. Mottura, and S. Gentili. 2004. Fossil *Macaca* remains from RDB quarry (Villafranca d'Asti, Italy): New data and overview. *Journal of Human Evolution* 40(3): 187–202.

Roonwal, M. L., and S. M. Mohnot. 1977. *Primates of South Asia: Ecology, sociobiology, and behavior.* Cambridge, MA: Harvard University Press.

Rosenberger, A. L., W. C. Hartwig, and R. G. Wolff. 1991. *Szalatavus attricuspis*, an early platyrrhine primate. *Folia Primatologia* 56(4): 225–233.

Rossan, R. N., and D. C. Baerg. 1975. Development of *falciparum* malaria in a Panamanian subspecies of howler monkey. *American Journal of Tropical Medicine and Hygiene* 24(6): 1035–1036.

Rossan, R. N., D. C. Baerg, and M. D. Young. 1975. Five species of Panamanian monkeys as new experimental hosts for *Plasmodium simium*. *Journal of Parasitology* 61(4): 768–769.

Rossignol, P. A., J. M. C. Ribeiro, and A. Spielman. 1984. Increased intradermal probing time in sporozoite-infected mosquitoes. *American Journal of Tropical Medicine and Hygiene* 33(1): 17–20.

Rothe, W. E., D. P. Jacobus, and W. G. Walter. 1969. Treatment of trophozoite-induced *Plasmodium knowlesi* infection in the Rhesus monkeys with trimethoprim and sulfalene. *American Journal of Tropical Medicine and Hygiene* 18(4): 491–494.

Roy, S. W., and M. Irimia. 2008. Origins of human malaria: Rare genomic changes and full mitochondrial genomes confirm the relationship of *Plamodium falciparum* to other mammalian parasites but complicate the origins of *Plasmodium vivax*. *Molecular Biology and Evolution* 25(6): 1192–1198.

Rueda, L. M., T. L. Brown, H. C. Kim, S. Chong, T. A. Klein, D. H. Foley, A. Anyamba, M. Smith, E. P. Pak, and R. C. Wilkerson. 2010. Species composition, larval habitats, seasonal occurrence and distribution of potential malaria vectors and associated species of *Anopheles* (Diptera: Culicidae) from the Republic of Korea. *Malaria Journal 9:55, www.malariajournal.com/content/9/1/55* (accessed December 3, 2010).

Ruvolo, M. 1997. Molecular phylogeny of the hominoids: Inferences from multiple independent DNA sequence data sets. *Molecular Biology and Evolution* 14(3): 248–265.

Ruwende, C., and A. Hill. 1998. Glucose-6-phosphate dehydrogenase deficiency and malaria. *Journal of Molecular Medicine* 76(8): 581–588.

Ruwende, C., S. C. Khoo, R. W. Snow, S. N. R. Yates, D. Kwiatokowski, S. Gupta, P. Warn, C. E. M. Allsopp, S. C. Gilbert, N. Peschu, C. I. Newbold, B. M. Greenwood, K. Marsh, and A. V. S. Hill. 2002. Natural selection of hemi- and heterozygotes for G6PD deficiency in Africa by resistance to severe malaria. *Nature* 376: 246–249.

Sabin, A. B. 1985. Oral poliovirus vaccine: History of its development and use and current challenge to eliminate poliomyelitis from the world. *Journal of Infectious Diseases* 151(3): 420–436.

Saenz, F., B. Balu, J. Smith, S. R. Menconca, and J. H. Adams. 2008. The transmembrane isoform of *Plasmodium falciparum* MAEBL is essential for the invasion of

Anopheles salivary glands. *PloS ONE* 3(5), *www.lbl.uab.edu:15128/picrender.fcgi?tool=pmcentrez&artid=2386256&blobtype=pdf* (accessed May 31, 2008).

Sallares, R., A. Bouwman, and C. Anderung. 2004. The spread of malaria to southern Europe in antiquity: New approaches to old problems. *Medical History* 48(3): 311–328.

Savolainen, P., Y. Zhang, J. Luo, J. Lundeberg, and T. Leitner. 2002. Genetic evidence for an East Asian origin of domestic dogs. *Science* 298(5598): 1610–1613.

Schall, J. J., and S. P. Vogt. 1993. Distribution of malaria in *Anolis* lizards of the Luquillo Forest, Puerto Rico: Implications for host community ecology. *Biotropica* 25(2): 229–235.

Schlagenhauf, P. 2004. Malaria: From prehistory to present. *Infectious Disease Clinics of North America* 204(2): 189–205.

Schmidt, L. H., Donald V. Cramer, Richard N. Rossan, and Janet Harrison. 1977a. The characteristics of *Plasmodium cynomolgi* infections in various Old World primates. *American Journal of Tropical Medicine and Hygiene* 26(3): 356–372.

———. 1977b. Differences in the virulence of *Plasmodium knowlesi* for *Macaca Irus* (*Fascicularis*) of Philippine and Malayan origins. *American Journal of Tropical Medicine and Hygiene* 26(4): 612–622.

Schmidt, L. H., R. Fradkin, J. Harrison, R. N. Rossan, and W. Squires. 1980. The course of untreated *Plasmodium inui* infections in the Rhesus monkey (*Macaca Mulatta*). *American Journal of Tropical Medicine and Hygiene* 29(2): 158–169.

Schmidt, L. H., R. Greenland, and C. S. Genther. 1961. The transmission of *Plasmodium cynomolgi* to man. *American Journal of Tropical Medicine and Hygiene* 10(5): 679–688.

Schmidt, L. H., R. Greenland, R. Rossan, and C. Genther. 1961. Natural occurrence of malaria in Rhesus monkeys. *Science* 133(3455): 753.

Schwartz, M. 1997. *A history of dogs in the early Americas*. New Haven, CT: Yale University Press.

Scopel, K. K. G., C. J. F. Fontes, A. C. Nunes, M. F. Horta, and É. M. Braga. 2004. High prevalence of *Plasmodium malariae* infections in a Brazilian Amazon endemic area (Apiacás—Mato Grosso State) as detected by polymerase chain reaction. *Acta Tropica* 90: 61–64.

Seethamchai, S., C. Putaporntip, S. Malaivijitnond, L. Cui, and S. Jongwuitiwes. 2008. Malaria and *Hepatocystis* species in wild macaques, southern Thailand. *American Journal of Tropical Medicine and Hygiene* 78(4): 646–653.

Sein, K. K., A. E. Brown, Y. Maeno, C. D. Smith, K. D. Corcoran, P. Hansukjariya, H. K. Webster, and M. Aikawa. 1993. Sequestration pattern of parasitized erythrocytes in cerebrum, mid-brain, and cerebellum of *Plasmodium coatneyi*-infected Rhesus monkeys (*Macaca mulatta*). *American Journal of Tropical Medicine and Hygiene* 49(4): 513–519.

Senga, E., M.-P. Loscertales, K. E. B Makwakwa, G. N. Liomba, C. Dzmalala, P. N. Kazembe, and B. J. Brabin. 2007. ABO blood group phenotypes influence parity specific immunity to *Plasmodium falciparum* malaria in Malawian women. *Malaria Journal* 6: 102, *www.malariajournal.com/content/pdf/1475-2875-6-102.pdf* (accessed May 30, 2008).

Service, M. W. 1989. Rice, a challenge to health. *Parasitology Today* 5(5): 162–165.

Shen, G., X. Gau, B. Gao, and D. E. Granger. 2009. Age of Zhoukoudian *Homo erectus* determined with $^{26}Al/^{10}Be$ burial dating. *Nature* 458: 198–200.

Sherman, I. W. 2005. The life cycle of *Plasmodium*: An overview. In *Molecular Approaches to Malaria*, I. W. Sherman, Ed., 3–11. Washington, D.C.: American Society for Microbiology Press.

Siddall, M. E., and J. R. Barta. 1992. Phylogeny of *Plasmodium* species: Estimation and inference. *Journal of Parasitology* 78(3): 567–568.

Silva, J. C., A. Egan, R. Friedman, J. B. Munro, J. M. Carlton, and A. L. Hughes. 2010. Genome sequences reveal divergent times of malaria parasite lineages. *Parasitology*, Pre-print on-line publication: DOI: 10.1017/S0031182010001575.

Simons, E. L., E. R. Seiffert, T. M. Ryan, and Y. Attia. 2007. A remarkable female cranium of early Oligogene *Aegyptopithecus zeuxis* (Catarrhini, Propliopithecidae). *Proceedings of the National Academy of Sciences, United States of America* 104(21): 8731–8736.

Sinden, R. E., Elizabeth U. Canning, R. S. Bray, and M. E. Smalley. 1978. Gametocyte and gamete development in *Plasmodium falciparum. Proceedings of the Royal Society of London. B.* 201: 375–399.

Sinden, R. E., and H. M. Gilles. 2002. The malarial parasites. In *Essential malariology,* fourth edition, D. A. Warell and H. M. Gilles, Eds., 8–34. New York: Oxford University Press.

Singh, B., and P. C. S. Divis. 2009. Orangutans not infected with *Plasmodium vivax* or *P. cynomolgi*, Indonesia. *Emerging Infectious Diseases* 15(10) (October), *www.cdc.gov/eid/content/15/10/1657.htm* (accessed December 21, 2009).

Singh, B., L. K. Sung, A. Matusop, A. Radhakrishnan, S. S. G. Shamsul BSc, J. Cox-Singh, A. Thomas, and D. J Conway. 2004. A large focus of naturally acquired *Plasmodium knowlesi* infections in human beings. *Lancet* 363(9414): 1017–1024.

Sinton, J. A., E. L. Hutton, and P. G. Shuts. 1938. Failure to transmit an infection of *P. cynomolgi* to man by blood injection and by mosquito bites. *Journal of Tropical Medicine and Hygiene* 41: 245–246.

Slater, L. B. 2005. A history of malaria research at NIAID. Paper presented as Dewitt Stetton lecture, November 8, in Bethesda, MD.

Small, M. F. 1994. Macaque see, macaque do. *Natural History* 103: 8–11.

Smith, C. D., A. E. Brown, S. Nakazawa, H. Fujioka, and M. Aikawa. 1996. Multi-organ erythrocyte sequestration and ligand expression in Rhesus monkeys infected with *Plasmodium coatneyi* malaria. *American Journal of Tropical Medicine and Hygiene* 55(4): 379–383.

Smith, D. G., J. W. Mcdonough, and D. A. George. 2006. Mitochondrial DNA variation within and among regional populations of longtail macaques (*Macaca fascicularis*) in relation to other species of the *fascicularis* group of macaques. *American Journal of Primatology* 69(2): 182–198.

Smith, N. J. H. 1981. Colonization lessons from a tropical forest. *Science* 214(4522): 755–761.

Snelders, S., C. Kaplan, and T. Pieters. 2006. On cannabis, chloral hydrate, and career cycles of psychotropic drugs in medicine. *Bulletin of the History of Medicine* 80(1): 95–114.

Snow, R. W., C. A. Guerra, A. M. Noor, H. Y. Myint, and S. I. Hay. 2005. The global distribution of clinical episodes of *Plasmodium falciparum* malaria. *Nature* 434: 214–217.

Solomon, E. P., L. R. Berg, and D. W. Martin. 2008. *Biology.* Eighth edition. Belmont, CA: Thomson Books.

Southwick, C. H., and F. F. Siddiqi. 1977. Population dynamics of Rhesus Monkeys in Northern India. In *Primate conservation*, Eds. His Serene Highness Prince Rainier III of Monaco and G. H. Bourne, 339–362. New York: Academic Press.

Southwick, C. H., M. F. Siddiqi, and J. R. Oppenheimer. 1983. Twenty-year changes in Rhesus monkey populations in agricultural areas of Northern India. *Ecology* 64(3): 434–439.

Southwick, C. H., Z. Yongzu, H. Haisheng, L. Zhenhe, Q. Wenyuan. 1996. Population ecology of Rhesus macaques in tropical and temperate habitats in China. In *Evolution and ecology of macaque societies*, J. E. Fa and D. G. Lindburg, Eds., 95–105. Cambridge: Cambridge University Press.

Spassov, N., and D. Geraads. 2007. *Dolichopithecus balcanicus* sp. nov., a new Colobinae (Primates, Cercopithecidae) from the early Pliocene of southeastern Europe, with a discussion on the taxonomy of the genus. *Journal of Human Evolution* 52(4): 434–442.

Stannard, J. 1993. Diseases of western antiquity. In *The Cambridge World History of Human Disease*, K. F. Kiple, Ed., 262–270. Cambridge: Cambridge University Press.

Steiper, M. E., N. D. Wolfe, W. B. Karesh, A. M. Kilbourn, E. J. Bosi, and M. Ruvolo. 2005. The population genetics of the α-2 Globin Locus of Orangutans (*Pongo pygmaeus*). *Journal of Molecular Evolution* 60(3): 400–408.

Steketee, R. W., B. L. Nahlen, M. E. Parise, and C. Menendez. 2001. The burden of malaria in pregnancy in malaria-endemic areas. *American Journal of Tropical Medicine and Hygiene* 64(1/2) S: 28–35.

Sterling, C. R., T. M. Seed, M. Aikawa, and J. Rabbege. 1975. Erythrocyte membrane alterations induced by *Plamodium simium* infection in *Saimiri sciureus*: Relation to Schüffner's dots. *Journal of Parasitology* 61(2): 177–188.

Storey, A. A., J. M. Ramírez, D. Quiroz, D. V. Burley, D. J. Addison, R. Walter, A. J. Anderson, T. L. Hunt, J. S. Athens, L. Huynen, and E. A. Matisoo-Smith. 2007. Radiocarbon and DNA evidence for a Pre-Columbian introduction of Polynesian chickens. *Proceedings of the National Academy of Science, United States of America* 104(25): 10335–10339.

Stuart, M., V. Pendergast, S. Rumfelt, S. Pierberg, L. Greenspan, K. Glander, and M. Clarke. 1998. Parasites of wild Howlers (*Alouatta spp.*). *International Journal of Primatology* 19(3): 493–512.

Sturm, A., R. Amino, C. van de Sand, T. Regen, S. Retzlaff, A. Rennenberg, A. Krueger, J.-M. Pollok, R. Menard, and V. T. Heussler. 2006. Manipulation of host hepatocytes by the malaria parasite for delivery into liver sinusoids. *Science* 313(5791): 1287–1290.

Sturm, A., and V. Heussler. 2007. Live and let die: Manipulation of host hepatocytes by exoerythrocytic *Plasmodium* parasites. *Medical Microbiology and Immunology* 196(3): 127–133.

Su, X.-Z., J. Mu, and D. A. Joy. 2003. The "Malaria's Eve" hypothesis and the debate concerning the origin of the human malaria parasite *Plasmodium falciparum*. *Microbes and Infection* 5(10): 891–896.

Sugiyama, M., E. Ikeda, S. Kawai, T. Higuchi, H. Zhang, N. Khan, K. Tomiyoshi, T. Inoue, H. Yamaguchi, K. Katakura, K. Endo, and M. Suzuki. 2004. Cerebral metabolic reduction in severe malaria: Fluorodeoxyglucose-positron emission tomography imaging in a primate model of severe human malaria with cerebral involvement. *American Journal of Tropical Medicine and Hygiene* 71(5): 542–545.

Sullivan, J., A. Bounngaseng, A. Stewart, J. J. Sullivan, G. G. Galland, F. Henry, and W. E. Collins. 2005. Infection of *Saimiri boliviensis* monkeys with *Plasmodium coatneyi*. *Journal of Parasitology* 91(2): 479–481.

Sullivan, J., D. Nace, T. Williams, J. Guarner, G. S. Noland, and W. E. Collins. 2003. The development of exoerthrocytic states of *Plasmodium inui shortii* in New World monkeys. *Journal of Parasitology* 89(3): 637–639.

Sullivan, J., A. Stewart, A. Bounngaseng, J. J. Sullivan, D. Nace, A. Williams, G. G. Galland, T. Williams, F. Henry, and W. E. Collins. 2006. Observations on the exoerythrocytic states of different isolates of *Plasmodium cynomolgi* in hepatocytes of New World *Aotus* and *Saimiri* monkeys. *Journal of Parasitology* 92(1): 202–205.

Sullivan, J., E. Strobert, C. Yang, C. L. Morris, G. G. Galland, B. B. Richardson, A. Bounngaseng, J. Kendall, H. McClure, and W. E. Collins. 2001. Adaptation of a strain of *Plasmodium vivax* from India to New World monkeys, chimpanzees, and anophiline mosquitoes. *Journal of Parasitology* 87(6): 1398–1403.

Sullivan, J. S., C. L. Morris, H. M. McClure, E. Strobert, B. B. Richardson, G. G. Galland, I. F. Goldman, and W. E. Collins. 1996a. *Plasmodium vivax* infections in chimpanzees for sporozoite challenge studies in monkeys. *American Journal of Tropical Medicine and Hygiene* 55(3): 344–349.

Sullivan, J. S., C. L. Morris, B. B. Richardson, G. G. Galland, and W. E. Collins. 1998. Attempts to transmit the N-3 strain of *Plasmodium fieldi* to *Aotus* monkeys. *Journal of Parasitology* 84(1): 195–197.

Sullivan, J. S., C. L. Morris, B. B. Richardson, G. G. Galland, V. M. Jennings, J. Kendall, and W. E. Collins. 1999. Adaptation of the AMRU-1 strain of *Plasmodium vivax* to *Aotus* and *Saimiri* monkeys and to four species of anopheline mosquitoes. *Journal of Parasitology* 85(4): 672–677.

Sullivan, J. S., C. L. Morris, B. B. Richardson, G. G. Galland, J. J. Sullivan, and W. E. Collins. 1996b. Sporozoite transmission of three strains of *Plasmodium knowlesi* to *Aotus* and *Saimiri* monkeys. *Journal of Parasitology* 82(2): 268–271.

Sullivan, J. S., V. M. Jennings, J. Guarner, G. S. Noland, J. Kendall, and W. E. Collins. 2002. Infection of *Aotus* and *Saimiri* monkeys with *Plasmodium gonderi*. *Journal of Parasitology* 88(2): 422–425.

Suzuki, M. 2004. Cerebral metabolic reduction in severe malaria: Fluorodeoxyglucose-positron emission tomography imaging in a primate model of severe human malaria with cerebral involvement. *American Journal of Tropical Medicine and Hygiene* 71(5): 542–545.

Swisher III, C. C., G. H. Curtis, T. Jacob, A. G. Getty, A. Suprijo, and Widiasmoro. 1994. Age of the earliest known hominids in Java, Indonesia. *Science* 263(5150): 1118–1121.

Swisher III, C. C., W. J. Rink, S. C. Antón, H. P. Schwarcz, G. H. Curtis, A. Suprijo, and Widiasmoro. 1996. Latest *Homo erectus* of Java: Potential contemporaneity with *Homo sapiens* in Southeast Asia. *Science* 274(5294): 1870–1874.

Tadei, W. P., B. D. Thatcher, J. M. M. Santos, V. M. Scarpassa, I. B. Rodrigues, and M. S. Rafael. 1998. Ecologic observations on anopheline vectors of malaria in the Brazilian Amazon. *American Journal of Tropical Medicine and Hygiene* 59(2): 325–335.

Takai, M., F. Anaya, N. Shigehara, and T. Setoguchi. 2000. New fossil materials of the earliest *Branisella boliviana* and the problem of platyrrhine origins. *American Journal of Physical Anthropology* 111(2): 263–281.

Taliaferro, W. H., and L. Graves Taliaferro. 1934. The transmission of *Plasmodium falciparum* to the howler monkey, *Alouatta* sp. I. General nature of the infections and morphology of the parasites. *American Journal of Hygiene* 19: 318–334.

Taylor, D. W., R. A. Wells, A. Vernes, Y. J. Rosenberg, S. Vogel, and C. L. Diggs. 1985. Parasitologic and immunologic studies of experimental *Plasmodium falciparum* infection in nonsplenectomized chimpanzees (*Pan troglodytes*). *American Journal of Tropical Medicine and Hygiene* 34(1): 36–44.

Telford, M. J., and R. R. Copley. 2005. Animal phylogeny: Fatal attraction. *Current Biology* 15(8): R296–R299.

Tobie, J. E., D. C. Abele, G. J. Hill II, P. G. Contacos, and C. B. Evans. 1966. Fluorescent antibody studies on the immune response in sporozoite-induced and blood-induced *vivax* malaria and the relationship of the antibody production to parasatemia. *American Journal of Tropical Medicine and Hygiene* 15(5): 676–683.

Tobie, J. E., and G. R. Coatney. 1964. The antibody response in volunteers with *cynomolgi* malaria infections. *American Journal of Tropical Medicine and Hygiene* 13(6): 786–789.

Tobie, J. E., S. F. Kuvin, P. G. Contacos, G. R. Coatney, and C. B. Evans. 1962. Fluorescent antibody studies on cross reactions between human and simian malaria in normal volunteers. *American Journal of Tropical Medicine and Hygiene* 11(5): 589–596.

Tongren, J. E., C. Yang, W. E. Collins, J. S. Sullivan, A. A. Lal, and L. Xiao. 2000. Expression of proinflammatory cytokines in four regions of the brain in *Macque mulatta* (Rhesus) monkeys infected with *Plasmodium coatneyi*. *American Journal of Tropical Medicine and Hygiene* 62(4): 530–534.

Trampuz, A., M. Jereb, I. Muzlovic, and R. M Prabhu. 2003. Clinical review: Severe malaria. *Critical Care* 7(4): 315–323.

Trevelyan, B., M. Smallman-Raynor, and A. D. Cliff. 2005. The spatial dynamics of poliomyelitis in the United States: From epidemic emergence to vaccine-induced retreat, 1910–1971. *Annals of the Association of American Geographers* 95(2): 269–293.

Trinkaus, E., O. Moldovan ştefan Milota, A. Bîlgăr, L. Sarcina, S. Athreya, S. E. Bailey, R. Rodrigo, G. Mircea, T. Higham, C. B. Ramsey, and J. van der Plicht. 2003. An early modern human from the Peştera cu Oase, Romania. *Proceedings of the National Academy of Science, United States of America* 100(20): 11231–11236.

Tsoukala, E., and A. Bartsiokas. 2008. New *Mesopithecus pentelicus* specimens from Kryopigi, Macedonia, Greece. *Journal of Human Evolution* 54(3): 448–451.

Udomsangpetch, R., A. E. Brown, C. D. Smigh, and H. K. Webster. 1991. Rosette formation by *Plasmodium coatneyi*-infected red blood cells. *American Journal of Tropical Medicine and Hygiene* 44(4): 399–401.

———. 1993. *Plasmodium coatneyi* ring-infected erythrocyte surface antigens. *American Journal of Tropical Medicine and Hygiene* 49(1): 127–134.

van Hellemond, J. J., M. Rutten, R. Koelewijn, A.-M. Zeeman, J. J. Verweij, P. J. Wismans, C. H. Kocken, and P. J. J. van Genderen. 2009. Human *Plasmodium knowlesi* infection detected by rapid diagnostic tests for malaria. *Emerging Infectious Diseases* 15(9): 1478–1480.

Varki, A. 2001. Loss of N-Glycolylneuraminic acid in humans: Mechanisms, consequences, and implication for hominid evolution. *Yearbook of Physical Anthropology* 44(S33): 54–69.

Verrelli, B. C., S. A. Tishkoff, A. C. Stone, and J. W. Touchman. 2006. Contrasting histories of G6PD molecular evolution and malarial resistance in humans and chimpanzees. *Molecular Biology and Evolution* 23(8): 1592–1601.

Vichinsky, E. P., E. A. MacKlin, J. S. Waye, F. Lorey, and N. F. Olivieri. 2005. Changes in the epidemiology of thalassemia in North America: A new minority disease. *Pediatrics* 111(6): e818–e825.

Volkman, S. K., A. E. Barry, E. J. Lyons, K. M. Nielsen, S. M. Thomas, M. Choi, S. S. Thakore, K. P. Day, D. F. Wirth, and D. L. Hartl. 2001. Recent origin of *Plasmodium falciparum* from a single progenitor. *Science* 293(5529): 482–484.

Volney, B., J.-F. Pouliquen, B. de Thoisy, and T. Fandeur. 2002. A sero-epidemiological study of malaria in human and monkey populations in French Guiana. *Acta Tropica* 82: 11–13.

Vululu, J. M., R. F. Beach, F. K. Atieli, J. C. McAllister, W. G. Brogdon, J. M. Roberts, R. W. Mwangi, and W. A. Hawley. 1999. Elevated oxidase and esterase levels associated with permethrin tolerance in *Anopheles gambiae* from Kenyan villages using permethrin-impregnated nets. *Medical and Veterinary Entomology* 13(3): 239–244.

Vululu, J. M., R. F. Beach, F. K. Atieli, J. M. Roberts, D. L. Mount, and R. W. Mwangi. 1994. Reduced susceptibility of *Anopheles gambiae* to permethrin associated with the use of permethrin-impregnated bednets and curtains in Kenya. *Medical and Veterinary Entomology* 8(1): 71–75.

Vythilingam, I., Y. M. NoorAzian, T. C. Huat, A. I. Jiram, Y. M. Yusri, A. H. Azahari, I. NorParina, A. NoorRain, and S. LokmanHakim. 2008. *Plasmodium knowlesi* in humans, macaques, and mosquitoes in peninsular Malyasia. *Parasites and Vectors* 1 (August 19), *www.ncbi.nlm.nih.gov/pmc/articles/PMC2531168/pdf/1756-3305-1-26.pdf* (accessed October 16, 2009).

Wallis, J., and D. R. Lee. 1999. Primate conservation: The prevention of disease transmission. *International Journal of Primatology* 20: 803–826.

Ward, S., B. Brown, A. Hill, J. Kelley, and W. Downs. 1999. Equatorius: A new hominoid genus from the middle Miocene of Kenya. *Science* 285(5432): 1382–1386.

Warren, M., W. H. Cheong, H. K. Fredericks, and G. R. Coatney. 1970. Cycles of jungle malaria in west Malaysia. *American Journal of Tropical Medicine and Hygiene* 19(3): 383–393.

Warren, M., G. R. Coatney, and J. C. Skinner. 1966. *Plasmodium jefferyi* sp. n. from *Hylobates lar* in Malaya. *Journal of Parasitology* 52(1): 9–13.

Warren, M., J. C. Skinner, and E. Guinn. 1966. Biology of the simian malarias of Southeast Asia. I: Host cell preferences of young trophozoites of four species of *Plasmodium*. *Journal of Parasitology* 52(1): 14–16.

Warren, M., and R. H. Wharton. 1963. The vectors of simian malaria: Identity, biology, and geographical distribution. *Journal of Parasitology* 49(6): 892–904.

Waters, A. P., D. G. Higgins, and T. F. McCutchan. 1991. *Plasmodium falciparum* appears to have arisen as a result of lateral transfer between avian and human hosts. *Proceedings of the National Academy of Sciences, United States of America* 88(8): 3140–3144.

_____. 1993. Evolutionary relatedness of some models of Plasmodium. *Molecular Biology and Evolution* 10(4): 914–923.

Weatherall, D. J. 2008. Genetic variation and susceptibility to infection: The red blood cell and malaria. *British Journal of Haematology* 141(3): 276–286.

Weatherall, D. J., and J. B. Clegg. 2001. Inherited haemoglobin disorders: An increasing global health problem. *Bulletin of the World Health Organization* 79:8, *www.scielosp.*

org/scielo.php?script=sci_arttext&pid=S0042-96862001000800005&lng=en&nrm=iso&tlng=en (accessed on November 20, 2008).

Wekesa, J. W., R. S. Copeland, and R. W. Mwangi. 1992. Effects of *Plasmodium falciparum* on blood feeding behavior of naturally infected *Anopheles* mosquitoes in western Kenya. *American Journal of Tropical Medicine and Hygiene* 47(4): 484–488.

Wellde, B. T., A. J. Johnson, J. S. Williams, and E. H. Sadun. 1972. Experimental infection with *Plasmodium falciparum* in *Aotus* monkeys. I: Parsitologic, Hematologic, and Serum Biochemical Determinations. *American Journal of Tropical Medicine and Hygiene* 21(3): 260–271.

Weller, R. E. 1994. Infectious and non-infectious diseases of owl monkeys. In *Aotus: The owl monkey*, J. F. Baer, R. E. Weller, and I. Kakoma, Eds., 177–215. San Diego: Academic Press.

Westaway, K. E., M. J. Morwood, R. G. Roberts, A. D. Rodus, J.-x Zhao, P. Storm, F. Aziz, G. van den Bergh, P. Hadi, Jatmiko, and J. de Vos. 2007. Age and biostratigraphic significance of the Punung Rainforest fauna, east Java, Indonesia, and implications for *Pongo* and *Homo*. *Journal of Human Evolution* 53: 709–717.

Wheatley, B. P. 1980. Malaria as a possible selective factor in the speciation of macaques. *Journal of Mammology* 61(2): 307–311.

———. 1999. The sacred monkeys of Bali. Long Grove, IL: Waveland Press.

Wheatley, B., R. Stephenson, H. Kurashina, and K. March-Kautz. 2002. A cultural primatological study of *Macaca fascicularis* on Ngeaur Island, Republic of Palau. In *Primates face to face: The conservation implications of human-nonhuman primate interconnections*, Eds. A. Fuentes and L. D. Wolfe, 240–253. Cambridge: Cambridge University Press.

White, N. J. 2008. *Plasmodium knowlesi*: The fifth human malaria parasite. *Clinical Infectious Diseases* 46: 172–173.

———. 2003. Malaria. In *Manson's tropical diseases*, C. G. Cookk and A. Zumla, Eds., 1205–1295. Philadelphia: W. B. Saunders.

Whitely, H. E., J. I. Everitt, I. Kakoma, M. A. James, and M. Ristic. 1987. Pathologic changes associated with fatal *Plasmodium falciparum* infection in the bolivian Squirrel monkey (*Saimiri sciureus boliviensis*). *American Journal of Tropical Medicine and Hygiene* 37: 1–8.

Wiersch, S. C., W. A. Maier, and H. Kampen. 2005. *Plasmodium (Haemamoeba) cathemerium* gene sequences for phylogenetic analysis of malaria parasites. *Parasitology Research* 96(2): 90–94.

Wiesenfeld, S. L. 1967. Sickle-cell trait in human biological and cultural evolution. *Science* 157(3793): 1134–1140.

Willerson, D., Jr., K. H. Rieckmann, P. E. Carson, and H. Frischer. 1972a. Effects of minocycline against chloroquine-resistant *falciparum* malaria. *American Journal of Tropical Medicine and Hygiene* 21(6): 857–872.

Willerson, D., Jr., K. H. Rieckmann, L. Kass, P. E. Carson, H. Frischer, and J. E. Bowman. 1972b. The chemoprophylactic use of diformyl diaminodiphenyl sulfone (DFD) in *falciparum* malaria. *American Journal of Tropical Medicine and Hygiene* 2(2): 138–143.

Williams, T. N., K. Maitland, S. Bennett, M. Ganczakowski, T. E. A. Peto, C. I. Newbold, D. K. Bowden, D. J. Weatherall, and J. B. Clegg. 1996. High incidence of malaria in α-thalassaemic children. *Nature* 383: 522–525.

Wilson, A. C., and V. M. Sarich. 1969. A molecular time scale for human evolution. *Proceedings of the National Academy of Sciences, United States of America* 63(4): 1088–1093.

Wirsing, R. L. 1985. The health of traditional societies and the effects of acculturation. *Current Anthropology* 26: 303–322.

Wolfe, L. D. 1991. Macaques, pilgrims, and tourists re-visited. *National Geographic Research & Exploration* 7: 241.

———. 1992. Feeding habits of the Rhesus monkeys (*Macaca mulatta*) of Jaipur and Galta, India. *Human Evolution* 7: 43–54.

———. 2002. Rhesus macaques: A comparative study of two sites, Jaipur, India, and Silver Springs, Florida. In *Primates face to face: The conservation implications of human-nonhuman primate interconnections*, A. Fuentes and L. Wolfe, Eds., 310–330. Cambridge: Cambridge University Press.

Wolfe, N. D. 1999. Pathogen evolution and exchange in Bornean orangutans. Ph.D. dissertation. Harvard School of Public Health, Cambridge, MA.

Wolfe, N. D., W. B. Karesh, A. M. Kilbourn, J. Cox-Singh, E. Bosi, H. A. Rahman, A. T. Prosser, B. Singh, M. Andau, and A. Spielman. 2002. The impact of ecological conditions on the prevalence of malaria among orangutans. *Vector-Borne and Zoonotic Diseases* 2(2): 97–103.

Wood, C. S. 1975. New evidence for a late introduction of malaria into the New World. *Current Anthropology* 16: 93–104.

World Health Organization. 2005. *World Malaria Report* 2005. Prepared by Roll Back Malaria, World Health Organization, and Unicef. Geneva: World Health Organization.

———. 2006. WHO gives indoor use of DDT a clean bill of health for controlling malaria, *www.who.int/mediacentre/news/releases/2006/pr50/en/index.html* (accessed June 1, 2010).

———. 2008a. Measles Fact Sheet. *www.who.int/mediacentre/factsheets/fs286/en/* (accessed November 25, 2009).

———. 2008b. Poliomyelitis Fact Sheet. *www.who.int/mediacentre/factsheets/fs114/en/* (accessed November 25, 2009).

———. 2009. Malaria. *www.who.int/mediacentre/factsheets/fs094/en/* (accessed November 27, 2009).

———. 2010. Malaria. *www.who.int/mediacentre/factsheets/fs094/en/* (accessed June 1, 2010).

Wu, X., L. A. Schepartz, D. Fal, and W. Liu. 2006. Endocranial cast of Hexian *Homo erectus* from south China. *American Journal of Physical Anthropology* 130(4): 445–454.

Wu, X., L. A. Schepartz, and C. J. Norton. 2010. Morphological and morphometric analysis of variation in the Zhoukoudian *Homo erectus* brain endocasts. *Quaternary International* 211: 4–13.

Yang, C., L. Xiao, J. E. Tongren, J. Sullivan, A. A. Lal, and W. E. Collins. 1999. Cytokine production in Rhesus monkeys infected with *Plasmodium coatneyi*. *American Journal of Tropical Medicine and Hygiene* 61(2): 226–229.

Yewhalaw, D., W. Legesse, W. Van Bortel, S. Gebre-Selassie, H. Kloos, L. Duchateau, and N. Speybroeck. 2009. Malaria and water resource development: The case of Gilgel-Gibe hydroelectric dam in Ethiopia. *Malaria Journal* 8: 21, *www.malariajournal.com/content/8/1/21* (accessed August 4, 2010).

Yokoyama, Y., C. Falguères, F. Sémah, T. Jacob, and R. Grün. 2008. Gamma-ray spectrometric dating of late *Homo erectus* skulls from Ngandong and Sambungmacan, Central Java, Indonesia. *Journal of Human Evolution* 55(2): 274–277.

Young, M. D. 1957. The response of *Plasmodium malariae* infections to pyrimethamine (daraprim). *American Journal of Tropical Medicine and Hygiene* 6(2): 223–224.

Young, M. D., J. M. Ellis, and T. H. Stubbs. 1946. Studies on imported malarias. 5: transmission of foreign *Plasmodium vivax* by *Anopheles quadrimaculatus*. *American Journal of Tropical Medicine and Hygiene* s1-26(4): 477–482.

———. 1947. Some characteristics of *vivax* malaria induced in Neurosyphilitic patients. *American Journal of Tropical Medicine and Hygiene* s1-27(3): 585–596.

Young, M. D., and D. E. Eyles.1949. The efficacy of cholorquine, quniacrind, quinine, and totaquine in the treatment of *Plasmodium malariae* infections (Quartan malaria). *American Journal of Tropical Medicine and Hygiene* 28(1): 23–28.

Young, M. D., and D. V. Moore. 1961. Chloroquine resistance in *Plasmodium falciparum*. *American Journal of Tropical Medicine and Hygiene* 10(3): 317–320.

Young, M. D., J. A. Porter, Jr., and C. M. Johnson. 1966. *Plasmodium vivax* transmitted from man to monkey to man. *Science* 153(3739): 1006–1007.

Zilversmit, M., and D. L. Hartl. 2005. Evolutionary history and population genetics of human malaria parasites. In *Molecular approaches to malaria*, I. W. Sherman, Ed., 95–109 . Washington, D.C.: American Society for Microbiology Press.

Zimmerman, P. A., I. Woolley, G. L. Masinde, S. M. Miller, D. T. McNamara, F. Hazlett, C. S. Mgone, M. P. Alpers, B. Genton, G. A. Boatin, and J. W. Kazura. 1999. Emergence of $FY{*}A^{null}$ in a *Plasmodium vivax*-endemic region of Papua New Guinea. *Proceedings of the National Academy of Science, United States of America* 96(24): 13973–13977.

INDEX

About the Author

Loretta A. Cormier is associate professor of anthropology at the University of Alabama at Birmingham. She is author of *Kinship with Monkeys, the Guajá Foragers of Eastern Brazil* (Columbia University Press) and numerous articles in historical ecology and ethnoprimatology (human-nonhuman primate interactions).